Cellular Variation and Adaptation in Cancer

Cellular Variation and Adaptation in Cancer

Biological Basis and
Therapeutic Consequences

MICHAEL WOODRUFF
MD, DSc, MS (Melb.), FRCS (Engl. and Edinb.)
FRACS, FACS (Hon), FRCPE (Hon.), FRS

Oxford   New York   Tokyo
OXFORD UNIVERSITY PRESS
1990

ST. PHILIP'S COLLEGE LIBRARY

Oxford University Press, Walton Street, Oxford OX2 6DP
Oxford New York Toronto
Delhi Bombay Calcutta Madras Karachi
Petaling Jaya Singapore Hong Kong Tokyo
Nairobi Dar es Salaam Cape Town
Melbourne Auckland
and associated companies in
Berlin Ibadan

Oxford is a trade mark of Oxford University Press

Published in the United States
by Oxford University Press, New York

© Michael Woodruff, 1990

All rights reserved. No part of this publication may be reproduced,
stored in a retrieval system, or transmitted, in any form or by any means,
electronic, mechanical, photocopying, recording, or otherwise, without
the prior permission of Oxford University Press

British Library Cataloguing in Publication Data
Cellular variation and adaptation in cancer:
biological basis and therapeutic consequences.
1. Man. Cancer. Cells
I. Woodruff. Sir, Michael
616.99'407
ISBN 0-19-854254-2

Library of Congress Cataloging in Publication Data
Woodruff, Michael F. A., Sir.
Cellular variation and adaptation in cancer: biological basis and
therapeutic consequences/Michael F. A. Woodruff.
Includes bibliographical references.
1. Carcinogenesis. 2. Cancer cells—Variation. 3. Cancer cells—
Adaptation I. Title.
[DNLM: 1. Cell Transformation, Neoplastic—drug therapy. 2. Cell
Transformation, Neoplastic—pathology. 3. Neoplasm Metastasis—drug
therapy. 4. Neoplasm Metastasis—pathology. QZ 202 W893c]
RC268.5.W62 1990   616.99'4—dc19   89-16350   CIP
ISBN 0-19-854254-2

Set by Bath Press Datagraphics, Bath, Avon
Printed and bound in Great Britain by
Bookcraft (Bath) Ltd, Midsomer Norton

Preface

Why are current methods of treating cancer not more effective? In recent years there have indeed been some major advances, notably in the treatment of acute lymphoblastic leukaemia, Hodgkin's disease, malignant tumours of the testis, Wilms' tumour, and to a lesser extent osteogenic sarcoma; but, on the other hand, there has not been much improvement in the results of treating carcinoma of the stomach, small cell carcinoma of the lung, myelogenous leukaemia, carcinoma of the breast, and a host of other neoplasms.

The reasons for our treatment failures are multiple and complex. The realization that cancer often becomes generalized at an early stage provides part of the answer, and prompted the introduction of so-called adjuvant chemotherapy, i.e. the use of chemotherapy in addition to surgery in the management of carcinoma of the breast and other tumours, but, while this has proved to be of value in some categories of patient, the early high hopes which accompanied its introduction have not been fulfilled.

Two factors of critical importance are, I believe, the heterogeneous nature of cancer cell populations and their capacity to diversify, which lie at the heart of what Foulds, many years ago, called *progression*.

The last decade has seen the start of a deeper study, at both the cellular and molecular levels, of the origin of cancer cells, the ecology of cancer cell populations, and the mechanisms by which changes in such populations are brought about. The purpose of the book is to review this work and examine its therapeutic implications.

The problems to be discussed concern both clinicians and cancer biologists, and the book is addressed to workers in both fields.

It is difficult for people with heavy clinical commitments to gain a real insight into the revolution that is taking place in biology; it is also difficult for those working at the frontiers of knowledge to appreciate how cancer is perceived by clinicians, and the problems it poses for them. To borrow an analogy used by Bobrow in discussing the relationship between clinicians and geneticists, each group needs a user-friendly interface with the other's field, for without such interfaces 'even the most powerful analytical machines are underused, error prone or even incomprehensible' (Bobrow 1988). Moved by a fellow feeling—which, according to David Garrick, makes one wondrous kind—I have tried

tried hard to make the book as user-friendly as possible without dodging the difficulties.

A brief explanation of how the book came to be written may, perhaps, be in order. Until I retired from the University Chair of Surgery in Edinburgh my professional life was devoted to the practice of surgery, teaching, and, when time permitted, research in the fields of transplantation and oncology. Since then I have been able to undertake full time research in the Medical Research Council's Clinical and Population Cytogenetics Unit, through the kindness of the Director, Professor H. J. Evans, and with the financial support of the Council. During this time my work has been concerned with the study of cancer at the cellular level, with particular reference to the interaction of cancer and host, and tumour clonality, but I have also had a ringside seat from which to view the growth of molecular genetics and its application to cancer research, and the opportunity of getting to know some of the participants and entering into a dialogue with them.

Edinburgh
April 1989

M. F. A. W.

Acknowledgements

I am very conscious of the great debt I owe to Professor H. J. Evans and his colleagues in the Medical Research Council Human Genetics Unit (formerly the Clinical and Population Cytogenetics Unit), who encouraged and guided me during my 10 years of posthumous research; and to the Medical Research Council, whose financial support made this research possible. I am especially grateful to colleagues in the Human Genetics Unit, and also to Dr J. A. Ansell and Dr H. S. Micklem of the Edinburgh University Department of Zoology, who have collaborated with me in various experiments; to Ms Sheila Mould for the indefatigable way in which she has helped me to track down references in the Unit Library and elsewhere; and to Mr Norman Davidson for converting my rough sketches into the drawings reproduced in Figs 3 and 6.

I thank the publishers, editors, and authors listed below for permission to reproduce the following figures:

The Oxford University Press and Drs Wyke and Weiss for Figs 1 and 2, reproduced from *Cancer Surveys*, vol. 3; Academic Press Inc. for Fig. 3, reproduced from my article in *Advances in Cancer Research*, vol. 50; The University of Otago Medical School and Mr V. T. Pearse for the unpublished photograph reproduced in Fig. 4; The Royal College of Surgeons of England, Mr R. M. Kirk (Editor), and Mr D. C. Bodenham for Fig. 5, reproduced from *Annals of the Royal College of Surgeons of England*, vol. 43; and Dr Jack Cuzick for Figs 7 and 8, reproduced from *Cancer Treatment Reports*, vol. 71.

I am indebted to various people for helpful comments on the manuscript at various stages in its production, especially Professors H. J. Evans, R. A. Weiss, and K. D. Bagshawe; and to Dr Helen Stewart, Dr Maureen Roberts, and Sir Patrick Forrest for information about therapeutic trials and screening for breast cancer.

I am very grateful to the staff of Oxford University Press for friendly co-operation in the production of the book.

Last, but not least, my thanks to my wife—my sternest critic—for help in the preparation of the manuscript and her continued support and encouragement.

Contents

1 Diversity of cells in tumours   **1**
  1.1 Introduction   1
  1.2 Cellular composition of tumours   2
    1.2.1 Normal cells   3
    1.2.2 Neoplastic cells   3
    1.2.3 Partly transformed cells   4
    1.2.4 Revertant cells   4
    1.2.5 Hybrid cells   4
  1.3 Diversity of neoplastic cells   5
    1.3.1 Evidence of diversity   5
    1.3.2 Origin, maintenance, and consequences of diversity   6

2 Transformation, carcinogenesis, and progression   **8**
  2.1 Definitions   8
  2.2 Multistep nature of transformation and carcinogenesis   9
  2.3 Progression   10
  2.4 Genetics of transformation and carcinogenesis   11
    2.4.1 Oncogenes   11
    2.4.2 Tumour suppressor genes   19
    2.4.3 Modulator genes   23
  2.5 Environmental factors affecting transformation   23
  2.6 Is transformation reversible?   25
    2.6.1 *In vitro* evidence of reversion   25
    2.6.2 *In vivo* evidence of reversion   26
    2.6.3 Therapeutic implications   27

3 Clonal composition of tumours   **28**
  3.1 Definitions   28
  3.2 Evidence of monoclonality   29
    3.2.1 Clonal markers and their limitations   29
    3.2.2 Conclusions   33

3.3	Pleoclonal tumours		35
	3.3.1	How many clones?	35
	3.3.2	Spatial distribution of clones	36
	3.3.3	Does the clonal composition remain constant?	37
3.4	Tumour clonality in relation to carcinogenesis		37

4 Dynamic heterogeneity of tumour cell populations — 41

- 4.1 Ecological principles — 41
- 4.2 The genetic basis of phenotypic diversity — 41
  - 4.2.1 Normal development in eukaryotes — 42
  - 4.2.2 Generation of cellular diversity in tumours — 44
- 4.3 Cell cycle dynamics — 47
- 4.4 Dynamics of lineage heterogeneity — 47
  - 4.4.1 Influence of the host — 47
  - 4.4.2 Interactions between tumour cell subpopulations — 49
- 4.5 Tumour progression and regression — 52

5 Invasion and metastasis — 54

- 5.1 Analysis of invasive growth — 55
- 5.2 Routes of metastasis. Analysis of the process — 55
  - 5.2.1 Metastasis by lymphatics — 56
  - 5.2.2 Metastasis by the bloodstream — 56
  - 5.2.3 Metastasis by surface implantation — 60
- 5.3 Is there a metastatic phenotype? — 60
  - 5.3.1 Is metastasis stochastic or selective? — 60
  - 5.3.2 Genetic basis of metastatic ability — 62
- 5.4 Dormancy — 64
  - 5.4.1 Dormancy as a clinical phenomenon — 64
  - 5.4.2 Experimental study of dormancy — 66
  - 5.4.3 Nature of the phenomenon. Some unanswered questions — 67
- 5.5 Regression of metastases — 69

6 Limitations of current treatment — 71

- 6.1 Assessing the results of treatment — 71
- 6.2 Local treatment — 72
  - 6.2.1 Limitations of local treatment — 72
  - 6.2.2 Local treatment of breast cancer — 76
  - 6.2.3 Local treatment of other tumours — 78
- 6.3 Wide field irradiation — 79
- 6.4 Chemotherapy — 79
  - 6.4.1 Scope and limitations — 79
  - 6.4.2 The problems of selectivity. Minimizing complications — 80

		6.4.3 Drug resistance	86
		6.4.4 The risk of inducing new neoplasms	87
	6.5	Endocrinological procedures	88
		6.5.1 Treatment of advanced disease	88
		6.5.2 Endocrine adjuvant therapy	89
7	**Ways forward**		**92**
	7.1	Introduction	92
	7.2	Control of gene expression	94
		7.2.1 Oligonucleotide analogues	94
		7.2.2 Oligopeptides	95
	7.3	Antibody-directed attack on cancer cells	96
		7.3.1 Immunochemotherapy	96
	7.4	Other immunological procedures	99
		7.4.1 Introduction	99
		7.4.2 Adoptive immunization with LAK cells or TIL	100
	7.5	Use of growth factors to combat damage to normal tissues caused by cytotoxic drugs	103
	7.6	Anti-angioneogenesis	104
	7.7	The problem of occult residual metastatic cancer	104

Epilogue — 107
References — 109
Index — 131

1
Diversity of cells in tumours

1.1 Introduction

The appropriate generic term for the pathological entities with which this book is concerned is *neoplasm*; the primary subdivision of neoplasms is into those classed as *benign* and those classed as *malignant*.

What constitutes a neoplasm? Various definitions have been proposed; none is entirely satisfactory. Invasive growth, capacity to metastasize, and damage to the host that is independent of the site of the lesion, often suffice to characterize a neoplasm as malignant; but it is difficult to devise criteria by which to distinguish between benign neoplasms and non-neoplastic hyperplasia. Lesions regarded as benign neoplasms are typically localized and often solitary, while conditions such as fibroadenosis of the breast and nodular goitre, which are not regarded as neoplastic, are usually diffuse, but there are many exceptions. Thus neurofibromas of the skin and fibromyomas of the uterus are often multiple, and fibroadenomas of the breast are multiple occasionally; on the other hand, fibroadenosis of the breast may be localized, and nodular goitre sometimes takes the form of a solitary nodule. Another distinction that is sometimes drawn is that non-neoplastic hyperplasia is often attributable to endocrine stimulation whereas little or nothing is known about the aetiology of benign neoplasms except when, as in the case of the common wart, a virus is clearly involved. But ignorance of aetiology seems a poor basis on which to construct a classification.

Some lesions currently classed as benign neoplasms appear to represent a stage on the road to malignancy, and frequently become malignant unless removed or destroyed at an early stage. This is true, for example, of intestinal polyps in people with familial intestinal adeno-polyposis, some bladder papillomas, and so-called fetal adenoma of the thyroid. Other benign tumours, like fibromyomas of the uterus and ganglioneuroma, become malignant occasionally, but others again, like papillomas and neurofibromas of the skin, meningiomas, and the common hard fibroadenoma of the breast, do so rarely if ever. Conceivably, some neoplasms that have gone through all the stages of carcinogenesis behave as benign because they are held in check by host defence mechanisms, but this could be very difficult to prove.

Benign neoplasms have been much less thoroughly studied than malignant neoplasms, presumably because they pose less of a threat to patients. In the writer's view they are of great biological interest, but because our knowledge of them is so limited we shall, in this book, be concerned mainly with malignant neoplasms, and will use the term *neoplasm* without qualification in this sense.

The term *cancer* is commonly used as a synonym for malignant neoplasm, and will be used in this way in this book. It is still sometimes confined to malignant neoplasms of epithelial origin, but for these we shall use the term *carcinoma*. To be consistent, the term *carcinogenesis* should denote the development of a carcinoma but, in conformity with general usage, we shall use it in an extended sense to denote the process that culminates in the development of any kind of cancer.

Benign neoplasms, and many but by no means all malignant ones, present clinically as localized swellings that are visible, palpable, or detectable by radiography or various forms of scanning, and these are properly termed *tumours* or *solid neoplasms*. So prone is man, however, in Dryden's phrase, 'to torture one poor word ten thousand times', that *tumour* is very often used as a synonym for neoplasms in general, and in this book we have reluctantly accepted this convention.

There are such obvious phenotypic differences between tumours that arise in different tissues and organs, and also between tumours that arise in similar sites, that cancer has aptly been described as 'not one disease but many'. Moreover, tumours induced experimentally with the same agent at similar sites in isogeneic animals, or even in the same animal, may differ significantly. Thus fibrosarcomas induced in mice by subcutaneous injection of methylcholanthrene possess, as a rule, their own unique set of strong tumour associated transplantation antigens (TATA), as shown by the fact that injection of irradiated tumour cells to a new host protects it from subsequent challenge with viable cells of the same tumour, but not, as a rule, from challenge with viable cells of a different fibrosarcoma induced in the same way in the same, or an isogeneic, host (see Woodruff 1980, Chapter 5).

There is also heterogeneity within a single tumour, in respect of both structure and cellular composition.

1.2 Cellular composition of tumours

A tumour is a complex eco-system comprising cells of various kinds and extracellular matrix (ECM). The cell population includes neoplastic and normal cells, and perhaps also *partly transformed cells* that have taken one or more of the steps on the road to malignancy, *revertant cells* that have regained a non-malignant phenotype, and *hybrid cells* formed by the fusion of two cells of the same or different kinds.

1.2.1 Normal cells

The normal cells found in tumours include lymphocytes of various categories, plasma cells, polymorphonuclear leucocytes, macrophages, fibroblasts, and endothelial cells. Except for those in the last two categories, these cells reach the tumour via the bloodstream from the bone marrow and other parts of the reticuloendothelial system. Fibroblasts are usually of local origin, but there is some evidence (see Ansell *et al.* 1986) that fibroblast-like cells of bone marrow origin, presumably derived from blood monocytes, are also present in some tumours.

In all tumours, except perhaps at a very early stage, there is a supporting framework of vascular connective tissue which, together with the various normal cells, constitutes the *stroma* of the tumour. Moreover, in the leukaemias, the leukaemic cells in the blood are derived from stroma-containing tissue in the bone marrow and elsewhere. The acquisition of a vascularized stroma is a critical step in the development of a tumour, and there is good evidence that a substance secreted by the tumour, termed *tumour angiogenesis factor* (TAF), plays a key role in this process (see Folkman 1974).

It is often illuminating to think of cancer as a struggle for survival between a population of aberrant, aggressive neoplastic cells and the host, i.e. the patient or animal in which the tumour arises (Woodruff 1980, 1986*a*), but like all reductionist views this is incomplete. The vascular stroma of a tumour conveys nutrients and other substances needed by the tumour but may also convey a variety of substances that inhibit tumour growth. The role of inflammatory cells, and in particular of macrophages, in tumours is also ambivalent. On the one hand, activated macrophages appear to be selectively cytotoxic for tumour cells (see Woodruff 1980). On the other hand, non-activated macrophages may promote the growth of neoplastic cells both *in vitro* and *in vivo*. This has been demonstrated by depositing a mixture of macrophages and neoplastic cells on a disc of Millipore membrane and then either incubating the discs in tissue culture medium or implanting them in a suitable animal (Woodruff 1982). It seems likely that this effect is mediated by growth factors secreted by the macrophages.

1.2.2 Neoplastic cells

The process of transformation, which underlies the development of neoplastic cells, will be discussed in Chapter 2. What concerns us here is that the neoplastic cells in a tumour include proliferating cells in various stages of the cell cycle, so-called G_0 cells that have stopped cycling temporarily but may later re-enter the cycle, and dying or dead cells. Some, at least, of the dividing cells must have unlimited proliferative potential if the tumour is to continue to expand indefinitely, but there may also be proliferating non-self-renewing cells that are capable of undergoing only a limited number of cell divisions. The term *stem cell*, though often loosely used, should be reserved for the proliferating self-renewing cells. Should a descendant of a neoplastic cell become, for any reason, incapable of further division, it will be termed a *doomed cell*.

In some *in vitro* systems transformed cells communicate via gap junctions with each other but not with normal cells (reviewed in (rev.) Yamasaki 1986). Under certain conditions, as we shall see (Sections 2.6.1, p. 26 and 2.6.2), small molecules can also pass from normal to transformed cells.

1.2.3 Partly transformed cells

If transformation involves more than one step, as is usually the case (Chapter 2), a tumour might be expected to contain some partly transformed cells. The number will depend on how long such cells can survive, and the possibility of demonstrating them will depend on the availability of appropriate markers.

1.2.4 Revertant cells

There is evidence, which will be examined in detail later (Section 2.6), that fibroblasts transformed *in vitro* by polyoma virus or chemical carcinogens (Sachs 1974), fibrosarcoma cells (Sachs 1974, 1987*a*), and myeloid leukaemic cells (Sachs 1987*a*) can revert *in vitro* to a non-malignant phenotype. Such cells are not necessarily normal in all respects—they may, for example, be aneuploid—and they may undergo re-reversion to a malignant phenotype. There is also evidence (Section 2.6) that some malignant neoplastic cells can revert to a non-malignant phenotype *in vivo*. This is true not only of mouse teratocarcinoma cells transplanted to blastocysts, but also of neoplastic cells in some animal and human tumours that are *autochthonous*, i.e. growing in the host in which they have originated.

1.2.5 Hybrid cells

It is easy to make hybrid cells by fusing two tumour cells, or a tumour cell and a normal cell, *in vitro*, and it has been shown that hybridization may occur between transplanted tumour cells and host cells *in vivo* (Wiener *et al.* 1972). It seems reasonable to postulate that hybrid cells may also occur in autochthonous tumours.

Hybrids of tumour and normal cells made *in vitro* are, as a rule, not tumorigenic if they retain all, or nearly all, of the complete set of chromosomes of each cell, but may regain tumorigenicity as the result of selective chromosome loss (Harris 1971, 1985; Klein *et al.* 1973). It seems reasonable to postulate that the same rule holds good with hybrids formed *in vivo*. These phenomena will be considered further when we come to consider the genetic basis of cancer and, in particular, tumour suppressor genes (Section 2.4.2).

1.3 Diversity of neoplastic cells

1.3.1 Evidence of diversity

There is abundant evidence from many laboratories (rev. Dexter and Calabresi 1982; Woodruff 1983; Heppner 1984; Edwards 1985; Daar 1987) that the neoplastic cells of experimental tumours, and also of spontaneously developing animal and human tumours, are, as a general rule, markedly heterogeneous in respect of such diverse characters as morphology in tissue sections; morphology and growth characteristics in tissue culture; karyotype; tumorigenicity on transplantation; metastatic capacity as judged either by spontaneous metastasis from a primary tumour transplant or by lung colony formation after intravenous injection of tumour cells; sensitivity to cytotoxic drugs, irradiation, and hyperthermia; cell surface receptors for hormones, growth factors, and lectins; enzyme production; cell surface antigens and immunogenicity; sensitivity to host attack mediated by specific antibody plus complement, sensitized T-cells, macrophages or NK cells (Woodruff 1980, 1986*b*); and perhaps expression of particular oncogenes (Albino *et al.* 1984).

It is important, in this context, to distinguish between phenotypic characters of three kinds:

1. Characters that are exhibited only by doomed cells (Section 1.2.2).

2. Non-hereditary characters that are a function of the stage in the cell cycle reached by the cell, or are determined by features of the environment in which the cell happens to be—hypoxic cells, for example, may be insensitive to irradiation and to various cytotoxic drugs (Sections 6.2.1, 6.4.1, 7.1).

3. Characters that define cell *lineages*, i.e. subpopulations that are capable of breeding true, at least for a number of cell generations (Heppner 1984).

It seems possible—indeed, in the writer's view, likely—that some karyotypic abnormalities fall in the first category. Admittedly, there are many karyotypic abnormalities, for example the Philadelphia chromosome (Section 3.2.1, p. 31), that are clearly heritable, but are cells that are grossly polyploid capable of dividing, or are at least some of them doomed? The question merits more attention than it has yet received. For a start, it would be of interest to prepare cell suspensions from tumours that possess both highly polyploid and near-diploid cells, collect samples of each kind by running the cells through a fluorescence-activated cell sorter after staining them quantitatively for DNA with a vital dye such as Hoechst 33342, and assess their capacity to grow in tissue culture.

In this book we shall be concerned primarily with heterogeneity in respect of differences in cell lineage, and, following Heppner, will use the term heterogeneity, when applied to cell populations, in this restricted sense. It is important to note that heterogeneity of this kind is not confined to tumour cell populations; it occurs also in populations derived from normal epithelium (Heppner 1984; Edwards 1985), members of which may differ *inter alia* in their expression of various surface markers and in their sensitivity to carcinogens.

The availability of appropriate monoclonal antibodies (mAbs) has greatly facilitated the study of tumour cell heterogeneity in respect of cell surface markers, and, in particular, of cell surface antigens, including tumour-associated antigens (TAA) and antigens of the major histocompatibility complex (MHC).

It has been shown by studying sections of tumour tissue that many mAbs raised to epithelial tumours stain only some of the cells of a tumour. The distribution of these stained cells varies; they may be confined to one or more large areas, occur in many small patches, or be scattered diffusely throughout the tumour. Some mAbs, on the other hand, appear to recognize antigens common to all the tumour cells (see Edwards 1985; Daar 1987).

Since different antibodies may stain different cells, it is clear that the apparent heterogeneity is not just a patchy staining artefact caused, for example, by poor fixation. The possibility that it might be due to different antigens being expressed in different phases of the cell cycle has also been excluded by studying cells from rapidly dividing tumour cell lines in a fluorescence-activated cell sorter after staining them for DNA as well as for surface antigen. The interpretation of these experiments is complicated by the fact that cells entering mitosis must have twice as much surface antigen as cells that have just divided, and the possibility that antigen expression may be a function of the duration of the cell cycle, but when all due allowance has been made the conclusion still holds good (see Edwards 1985).

The number of distinct cell lineages in a tumour cell population is given by the number of different ways in which the various individual hereditary differences are combined. This number may well be large; Heppner (1984) has gone so far as to suggest that it may even reach the limiting case where 'the mosaic of characteristics of every isolated cell is different from that of every other.'

1.3.2 Origin, maintenance, and consequences of diversity

If a tumour originates from one normal cell, as appears usually to be the case (Chapter 3), heterogeneity must arise in partly or fully transformed cells; if, on the other hand, a tumour originates from two or more normal cells, heterogeneity may be due partly to heritable differences between these normal progenitors.

It is not surprising that a particular marker which is present in a normal progenitor cell should be conserved during transformation and be present in all the neoplastic cells which are descendants of this progenitor. It would, however, be remarkable if all the neoplastic cells carrying such a marker also carried a unique marker of a kind found only in neoplastic cells. Evidence that this sometimes happens, and possible explanations, will be considered later (Section 3.2.1, pp. 32–33) when we come to discuss tumour-associated antigens as markers of tumour clonality.

The characteristics of the neoplastic cell population in a tumour may change markedly, and sometimes suddenly. As a consequence of such changes, the tumour may behave in a more aggressive way, and this phenomenon is termed *progression*.

In analysing changes in the neoplastic cell population of a tumour we must consider (1) changes in genes and gene expression which alter the phenotypic features of individual cells; (2) short-term effects reflecting differences in how cells of different lineage respond to changes in the host environment; and (3) long-term changes in the population resulting from selective pressures occasioned by tumour–host interaction, competition between different tumour cell subpopulations, and treatment. Such analysis is facilitated by studying subpopulations obtained by cloning tumour cells *in vitro*, but this is apt to be misleading unless it is combined with the study of uncloned populations and whole tumours.

Heritable changes in neoplastic cells and the ecology of tumour cell populations form the subject of Chapter 4. The phenomenon of metastasis, the significance of the heterogeneity and adaptability of tumour cells as obstacles to the cure of cancer, and possible new therapeutic strategies, will be considered in Chapters 5, 6, and 7, respectively.

2
Transformation, carcinogenesis, and progression

2.1 Definitions

The neoplastic cells in a tumour have all arisen from one or more cells that have undergone a heritable process termed *neoplastic transformation* or, simply, *transformation*.

When transformed cells are grown in tissue culture they differ from normal cells in respect of various morphological and cell surface features, and proliferative potential. Unlike normal cells, which stop proliferating in monolayers when they become overcrowded and can be maintained in culture for only a limited number of cell generations, transformed cells are insensitive to overcrowding, as shown by their loss of contact and density-dependent inhibition of cell division, and may be maintained indefinitely in culture—a property termed *immortalization*.

Normal cells exposed *in vitro* to agents that cause transformation *in vivo* may assume all the characteristics of cultured tumour cells, in which case they too are said to be transformed. If, as may happen, they form a tumour when transplanted to a suitable host, they are said to be *tumorigenic* or *neoplastic*. The same terms are also used to describe cells that give rise to tumours in the autochthonous host. If cells transformed *in vitro* have not been tested for tumorigenicity, or if the test has proved negative, they may be described as having undergone *morphological transformation*, but not, of course, as tumorigenic.

Does the appearance of the *in vitro* features of transformation represent the same end-point as tumorigenicity? For the question to be meaningful it would be necessary for the conditions under which tumorigenicity is to be tested to be specified precisely, because the result will depend not only on intrinsic properties of the cells but also on the number of cells transplanted, the site of transplantation, and various features of the host environment.

The development of a tumour in an autochthonous host is similarly influenced by host factors, such as the local microenvironment, hormone levels, and host resistance mediated by immunological mechanisms, or by mechanisms of the kind that the writer has termed *para-immunological*, which are not necessarily triggered by antigen and are mediated by macrophages, NK cells, and possibly other cells, some of which may also play a role in mechanisms of a strictly immunological kind (Woodruff 1980). This makes it impossible, at a cellular level, to set an end point to carcinogenesis, or even to define precisely the number of steps involved (Section 2.2), because a cell that has taken enough steps to give rise to a tumour in a relatively favourable host environment might not have been able to do so in a less favourable environment until it had taken one or more additional steps.

Transformed cells tend to retain some of the lineage characters (Section 1.3.1) of the cells from which they have arisen. This is illustrated by the observation of Fiszman and Fuchs (1975) that when sarcoma cells that had been derived by transforming muscle cells with a Rous sarcoma virus that carried a temperature-sensitive oncogene, and which lacked the morphological features of differentiated muscle cells, were cultured for 24 hours at the non-permissive temperature, they began to form differentiated myotubes.

2.2 Multistep nature of transformation and carcinogenesis

It is widely accepted that transformation and carcinogenesis are multistep processes, except perhaps in the case of some animal tumours induced with retroviruses (Klein and Klein 1985).

There are four grounds for this belief:

1. Analysis of age incidence curves for human cancer points to multistep development, and many investigators have constructed multistep models to explain the epidemiological data (see Knudson 1973; Peto 1977). This evidence is not quite conclusive, however, because to construct such models assumptions that may not turn out to be true have to be made about various parameters that are difficult or impossible to observe and, as Peto has said, 'very similar predictions for the few things we can actually observe may follow from mathematical elaboration of very different multistage models'.

2. Cells that are abnormal histologically, karyotypically or functionally, but lack some of the hallmarks of fully transformed cells, are commonly found in close juxtaposition to cancer cells or at sites where cancer subsequently develops (see Farber and Cameron 1980). Such cells, which include some of the cells labelled *anaplastic* and *metaplastic* by pathologists, are a common feature of human cancer.

3. With some tumours, early changes appear to be brought about by agents known as *initiators* and later changes by agents of a different kind termed *promoters*. This was first demonstrated with experimentally induced skin cancers in mice by Berenblum, and has since been demonstrated with experimental tumours of various kinds (see Farber and Cameron 1980). With other tumours,

while initiation and promotion appear to be distinct processes, the same agent may act as both initiator and promoter (see Dulbecco 1982).

4. An oncogene (Section 2.4.1) which, when acting alone, does not transform normal cells, may do so in co-operation with a different oncogene, and may alone transform cells that have become *immortalized* but are not as yet fully transformed. Particular instances of this phenomenon will be discussed in Section 2.4.1, pp. 15-16.

When carcinogenesis involves two or more steps, the question arises of how to define a carcinogen. Some authors restrict this term to external agents that initiate carcinogenesis, but we will follow Peto (1977) and include agents that act, directly or indirectly, at any stage of the process. The qualification directly or indirectly allows, among other things, for the possibility that the agent acts by destabilizing the normal interaction between the cell and the extracellular matrix rather than on the cell itself (see Bissell *et al.* 1982; Bissell 1988; Simon-Assman *et al.* 1988). One might go further and omit the qualification *external*, so that inherited factors were also included, but to do this would be to depart too far from accepted usage. Instead, we shall make use of another term, the *oncogenic package*, to denote the whole system of genetic and environmental factors involved in the development of a tumour, and their complex interactions (Woodruff 1988).

In the case of animal tumours induced by directly-acting retroviruses, it can be argued that host susceptibility to the virus forms part of the oncogenic package. If so, these tumours can no longer be regarded as exceptions to the rule that carcinogenesis is a multistep process.

2.3 Progression

The concept of multistep carcinogenesis is linked with that of *progression*. This term was introduced by Rous and Beard (1935) to describe the metamorphosis of a virus-induced papilloma to a carcinoma, or more generally how 'tumours go from bad to worse', and subsequently defined by Foulds (see Foulds 1969) as 'the development of a tumour by way of permanent, irreversible qualitative change in one or more of the characters of its cells'. The question of whether transformation is irreversible will be discussed later (Section 2.6); meanwhile, it seems wise to omit Foulds' assumption of irreversibility, and to define progression simply as stepwise neoplastic development that may begin before transformation is complete and continue thereafter throughout the life of the host, or even beyond this if the tumour is transplanted to a new host.

Progression, and the mechanisms underlying it, will be discussed further in Chapter 4.

2.4 Genetics of transformation and carcinogenesis

The application of the methods of molecular genetics to the study of transformation—a growth industry if ever there was—is providing new insights into the complex mechanisms that regulate differentiation and maintain normal tissue equilibrium, and what happens when these mechanisms are deranged. This work has important implications for our understanding of carcinogenesis and of the generation of diversity among cancer cells, so we shall summarize the main conclusions that have emerged.

It now seems clear, as Klein and Klein (1985) have pointed out, that at least three functionally different groups of genes are involved in cancer: *oncogenes*, which, if activated, may bring about transformation; *tumour suppressor genes*, sometimes less happily called *anti-oncogenes*, which may prevent this; and *modulators*, which are not concerned in transformation but 'can affect the phenotypic properties of established tumours in a prognostically important way'.

2.4.1 Oncogenes

Definitions. Discovery of oncogenes

An *oncogene* is, strictly speaking, a gene whose expression plays an aetiological role in carcinogenesis; the term is, however, often used to include all genes which, when acting alone or in combination with other genes, can induce the transformation of cells in tissue culture.

Over 40 oncogenes have now been identified. About one quarter of these occur in the genomes of DNA viruses, such as Epstein–Barr (EB) virus, polyoma virus and SV40; the others originate in the genomes of cells but may have homologues in the genomes of RNA retroviruses.

Our present understanding of oncogenes stems from three discoveries (for review see Bishop 1981, 1987; Sattaur 1984; Balmain 1985; Klein and Klein 1985; Goyns 1986):

1. In tumorigenesis by the Rous sarcoma virus and other retroviruses a single gene determines the neoplastic phenotype.

In the case of the Rous sarcoma the gene is a hybrid between a cellular oncogene (*v. infra*) and a viral *env* gene, and its expression is influenced by the viral long terminal repeat (LTR) promoter and enhancer (see Marshall and Rigby 1984).

2. Homologues of the retroviral oncogenes exist in the genomes of normal animal and human cells. These homologues are termed *proto-oncogenes*. Viral oncogenes and their cellular homologues are distinguished by an appropriate prefix; thus v-*myc* is a viral oncogene of the *myc* family, and c-*myc* is its cellular homologue.

Some 20 retroviral oncogenes are now known, and of these at least 9 (v-*abl*, v-*erb B*, v-*ets*, v-*mos*, v-*myb*, v-*myc*, v-*H-ras*, v-*K-ras*, v-*sis*) have cellular counterparts that have been incriminated in tumorigenesis (Bishop 1987).

The discovery that RNA oncogenic viruses may be transmitted hereditarily, like cellular genes, through the ovum or sperm led Huebner and Todaro (1969) to postulate that 'the cells of many, and perhaps all, vertebrates contain information for producing C-type RNA viruses' and that 'the viral information (the virogene), including a portion responsible for transforming a normal cell into a tumor cell (the oncogene) is most commonly transmitted from animal to progeny animal and from cell to progeny cell in a covert form. Carcinogens, irradiation and the normal ageing process all favor the partial or complete activation of these genes'. It has been further postulated (Weiss 1973; Weiss et al. 1973; Fischinger and Haapals 1974), and is now generally accepted (see Hill and Hillova 1976; Todaro 1978), that the retroviral oncogenes have been acquired during evolution as the result of recombinatorial accidents between retroviruses and normal cellular genes (i.e. proto-oncogenes). It has been argued by Duesberg (1987) that this is an 'overinterpretation' of the observed homology, and that the cellular genes involved are not normal genes but, as we shall see (Section 2.4.1, p. 13), the fact that proto-oncogenes may be activated by chromosomal translocation tells strongly against this view.

3. The term proto-oncogene is also applied to similar cellular genes for which no viral homologue is known, including some members of the *ras* and *myc* families. These genes are sometimes present in an activated or mutated state in cells transformed chemically that are not virus infected, and in the cells of various animal and human tumours. This is shown by the fact that DNA from such cells, but not from normal cells, transforms cells of the NIH 3T3 line in culture by transfection. The 3T3 line was derived from mouse embryo fibroblasts; the cells are immortalized but not fully transformed.

Normal functions of proto-oncogenes

Proto-oncogenes code for growth factors, growth factor receptors, protein kinases (mostly tyrosine specific), cytoplasmic proteins that bind guanine triphosphate (GTP), and nuclear proteins that may modulate gene expression (rev. Sattaur 1984); they are thus involved in regulating cell growth and differentiation. Transfer of proto-oncogenes coding for growth factors, including a synthetic polynucleotide encoding epidermal growth factor (Stern et al. 1987), to normal cells may result in transformation; conceivably, as Monier (1987) has suggested, given the right recipient cells, any gene coding for a growth factor could behave as an oncogene.

It has been suggested by Weinberg (1985) that proto-oncogenes may also be involved in various cellular functions that are unrelated to growth and its regulation. This notion stems from the fact that in chickens, rats and Drosophila the *c-src* gene is expressed in high levels in cells of the nervous system, including fully differentiated neurons that do not divide, and may therefore be involved in some aspect of neuronal function. Conceivably, according to Weinberg, the *c-src* gene in its normal form is never involved in growth regulation, and 'its association with growth deregulation may be a consequence of the rare accident that caused its activation by a transducing retrovirus'.

Activation of proto-oncogenes

Proto-oncogenes may be activated (1) by integration of the genome of a retrovirus in the vicinity of the proto-oncogene, as a result of which the proto-oncogene becomes subject to regulation by viral promoter-enhancer segments; and (2) by non-viral mechanisms initiated by various agents such as chemical carcinogens and radiation.

Integration of a viral promoter-enhancer is the basis of the oncogenic effect of the slowly transforming retroviruses, such as the avian and rodent leukaemia viruses. Acutely transforming retroviruses, like the Rous and other sarcoma viruses, carry their own oncogene and act independently of the host genome. To explain the tumorigenicity of these derivatives of seemingly harmless cellular genes it has been suggested that they are driven by potent viral signals that the host cannot control, or that they have mutated while *en route* from proto-oncogene to oncogene (Bishop 1987). The conservation of these retroviral genes during evolution suggests that they may confer some advantages on the viruses that carry them.

Possible non-viral mechanisms include point mutation, DNA rearrangement, chromosomal translocation and gene amplification.

Point mutation and rearrangement lead to the production of a protein which differs from the normal gene product.

Chromosomal translocation involving an oncogene may simply increase the amount of a normal gene product, as is the case when c-*myc* comes under the influence of promoter-enhancer sequences from the immunoglobulin genes, or result in the production of a new protein. In the common 9/22 translocation responsible for the Ph^1 chromosome in chronic myeloid leukaemia, for example, the amino-terminus of the normal c-*abl* gene is translocated to a break point near the centre of the large *bcr* gene on chromosome 22, resulting in the formation of a *bcr*-c-*abl* fusion gene whose product is a protein of 210 kD molecular weight, instead of the normal 145 kD protein. When, as sometimes happens, the Ph^1 chromosome occurs in acute lymphoblastic leukaemia (ALL), it is again the result of a 9/22 translocation involving the *abl* gene, but now this gene is translocated to the 5' region of the *bcr* gene and the resulting product is a 190 kD protein (Hermans *et al.* 1987).

Gene amplification (see Alitalo and Schwab 1986; Evans 1986) results in increased production of a normal gene product. Very high amplification of an oncogene (up to several hundred-fold) has been observed in various human neoplasms and, as will be seen later (Section 4.2.2), this appears to be an important factor in tumour progression and in the response of tumours to environmental changes; lesser degrees of amplification occur earlier in the life of a tumour and may be involved in transformation.

How oncogenes transform cells

The contribution of a particular oncogene to transformation depends on the type of cell; but 'some oncogenes are more specific than others, and none are totipotent' (Klein and Klein 1986).

Oncogenes deregulate mechanisms concerned in normal cell growth and may block the differentiation of cells of a particular lineage at a particular stage of development. These effects may be produced in various ways (Fig. 1):

1. *The product of the oncogene is itself a growth factor (GF).*

This is exemplified by c-*cis*, which is an altered form of a gene whose normal function is to produce a subunit of platelet-derived growth factor (PDGF) (see Johnsson *et al*. 1984).

The GF may be adsorbed to appropriate receptors on the cell which secreted it or on another cell. Sporn and Todaro (1980) introduced the terms *autocrine stimulation* and *paracrine stimulation* to distinguish between these possibilities.

2. *The oncogene product is a normal or modified GF receptor.*

Modified receptors may bombard the cell with growth stimulating signals even when they are not stimulated by the corresponding GF (see Weinberg 1985). This possibility first became apparent when it was shown that the product of the v-*erb*-B oncogene resembles the receptor for epidermal growth factor (EGF) from which the EGF binding site has been deleted (Downward *et al*. 1984). It has been suggested that *src* and *abl* may act in a similar way.

3. *The oncogene product modifies transduction.*

The level or nature of cytoplasmic proteins that normally carry signals ('second messages') from stimulated GF receptors is altered in such a way that signals are received in the nucleus even when the receptor is not activated. Weinberg (1985) has suggested that proteins encoded by oncogenes of the *ras* family can act in this way.

4. *The oncogene product acts in the cell nucleus by deregulating the transcription of other genes.*

Oncogenes of the *myc* family act in this way.

One possibility is activation of a gene coding for a GF. Other possibilities are that the cell may be made independent of a GF that it normally requires, or becomes immortalized as the result of production of agents as yet unidentified.

Genetics of transformation and carcinogenesis 15

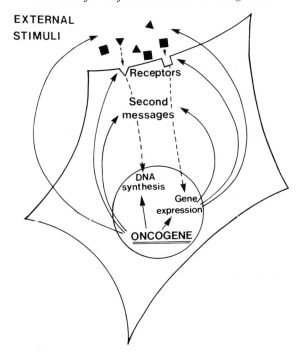

Fig. 1. Schematic representation of the ways in which untoward c-*onc* activity may disrupt the normal regulation of cell growth. The figure depicts a cell, with nucleus, that bears on its surface receptors (open square and triangle) for external stimuli such as growth factors (solid squares and triangles). Arrows on the left of the figure indicate that an oncogene in the cell DNA may directly encode molecules acting as external factors, receptors for these factors or components of second messages that lead to specific gene expression or DNA synthesis. Arrows within the nucleus indicate that DNA synthesis or specific gene expression may be mediated directly by the expression of an oncogene product located in the nucleus. Arrows on the right of the figure suggest that oncogene-induced gene expression may have an effect on reception and processing of environmental factors. (From Wyke, J. A. and Weiss, R. A. (1984). *Cancer Surveys*, **3**. By permission of Oxford University Press and the authors.)

It has been suggested by Land *et al.* (1983) that transformation, in so far as it is a multistep process (see Section 2.2), will require the co-operation of multiple oncogenes. This gains support from the observation that transfection with a human *ras* oncogene alone does not transform rodent fibroblasts in culture unless, like the cells of the NIH 3T3 line, they have already become immortalized; whereas co-transfection with a *ras* gene and another gene such as v-*myc*, c-myc, or the gene coding for the polyoma large-T antigen, does result in transformation.

The observation of Spandidos and Wilkie (1984) that transfection with a single *ras* gene that had been upregulated by genetic manipulation could bring about transformation of early passage rodent cells appears at first sight to be contrary to the prediction of Land *et al.*, but Spandidos and Wilkie found that the cells transfected in this way showed various chromosomal aberrations, and it is conceivable that these may have resulted in the activation of another gene or genes to complement the *ras* gene (see Balmain 1985).

The role of activated proto-oncogenes in carcinogenesis

The close relationship between proto-oncogenes and the acutely transforming oncogenes of retroviruses (Section 2.4.1, p. 13) suggests that proto-oncogenes are potential cancer genes, and that this potential may be realized when they are activated. Despite Rubin's (1984) objection that most of the reports of oncogene activity in tumour cell DNA have been based on studies using DNA from long-established tumour cell lines and immortalized NIH 3T3 cells as targets, it is now generally accepted that activated proto-oncogenes do play a major role in carcinogenesis, subject to two provisos: (1) two or more oncogenes acting synergistically, either sequentially or cumulatively but not necessarily in a particular order (Fig. 2), are likely to be needed, and (2) other events

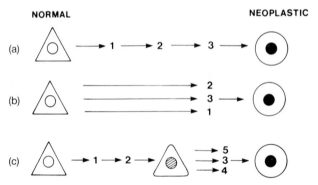

Fig. 2. Multiple steps in neoplasia. The numbers denote changes in gene function or expression that lead to neoplasia, and three modes are shown by which a normal cell may be converted into a tumour cell. (a) A sequential model in which the stages are in a fixed order and events important at any particular stage are only effective if the changes required for earlier stages have already occurred. (b) A cumulative model in which changes can occur in any order, neoplasia occurring when a number of changes have accumulated (this number may be less than the total number of possible events). (c) Sequential events convert a normal into a preoplastic cell which then progresses by cumulative changes in any order to a fully neoplastic condition. (From Wyke, J. A. and Weiss, R. A. (1984). *Cancer Surveys*, **3**. By permission of Oxford University Press and the authors.)

may also be necessary. The effect of an oncogene which perturbs the concentration of small molecules that act as second messengers in the growth response cascade may, for example, be so dampened by the leakage of such molecules to adjacent cells via gap junctions as to be ineffective (rev. Paul 1988), but Paul (1988) has suggested that, in solid tumours, this dampening may be reversed by another event which directly reduces the permeability of gap junctions or results in the formation of a critical mass of transformed cells.

To illustrate the effect of activated oncogenes in human tumours we will briefly review studies of Burkitt's lymphoma, chronic myeloid leukaemia, small cell carcinoma of the lung, and carcinoma of the colon and rectum.

Burkitt's lymphoma (BL). In African BL, as distinct from the sporadic form that occurs sometimes in other countries, the cells in 97 per cent of cases carry multiple copies of the Epstein–Barr Virus (EBV) genome. EBV is known to cause B-cell activation and immortalization *in vitro*, and to stimulate production of B-cell growth factor (BCGF), and it seems quite likely, therefore, that transformation of cells of B-lymphocyte lineage by EBV is an early event in the development of the tumour which makes the cells dependent on, and responsive to, BCGF. Chronic malaria, which appears to be aetiologically important, may act by increasing the size of the target cell population. In addition, c-*myc* is activated or dysregulated as the result of a chromosomal translocation that brings it into juxtaposition with one of the three immunoglobulin (Ig) loci that are normally active in B-lymphocytes. As a result, the *myc* gene, that is normally turned off in non-proliferating cells, becomes subject to *cis*-control by the constitutionally active Ig region and the cells continue to cycle, even when the waning of an antigenic stimulus would normally favour the transition of a clone of proliferating B cells to memory cells (Klein and Klein 1986). In some long-established Burkitt cell lines another c-*onc* gene, possibly a *ras* gene or *Blym*, is activated. This seems to account for the fact that DNA from these lines transforms NIH 3T3 cells, and it may well be that *in vivo* this gene acts in a complementary or synergistic way with c-*myc* (see Klein and Klein 1985).

Chronic myeloid leukaemia (CML). In the common translocation which gives rise to the Ph¹ chromosome seen in the great majority of patients with CML (Section 2.4.1, p. 13) the c-*abl* oncogene, located on the tip of chromosome 9, is transposed to chromosome 22, and the distal arm of chromosome 22 moves to chromosome 9. Less commonly, the deleted distal arm of chromosome 22 moves to an autosome other than chromosome 9, but it still happens that c-*abl* moves to chromosome 22. As Klein and Klein (1985) suggest, the consistency of the c-*abl* transposition greatly increases the probability that it is aetiologically significant in this disease.

Small cell carcinoma of the lung (SCLC). In most cases, as we shall see (Section 2.4.2, pp. 21–22), there is a characteristic deletion in the short arm of chromosome 3 that may result in the loss of both alleles of a tumour suppressor gene. In the present context the significant fact is that in about 20 per cent of cases of SCLC, in the neoplastic cells, or in cell lines derived from them, the c-*myc* and N-*myc* genes are found to be amplified 20–80-fold, and this change at the molecular level is associated with various phenotypic changes including increased cloning efficiency, more rapid growth, decreased expression of differentiation markers, and decreased adhesiveness *in vitro*, and a poorer prognosis, associated with a greater tendency to metastasize, *in vivo*. In some patients, cells of the variant population appear as a subpopulation in the primary tumour or in one or more metastases (Little *et al.* 1983).

Colorectal carcinomas. Hybridization assays of DNA extracted from colorectal carcinomas have revealed *ras* gene mutations, mostly in codon 12 of the c-Ki-*ras* gene (Bos *et al.* 1987). Assays of DNA from regions of tumour in which the tissue appeared histologically to be predominantly adenomatous pointed to the conclusion that *ras* mutation occurs also in colonic adenomas, which are generally accepted as being precancerous.

Forrester *et al.* (1987) reached similar conclusions by a different method of assay based on the ability of RNase A to recognize and cleave single base mismatches in RNA–RNA duplexes. They found mutated *ras* genes in many, but not all, carcinomas and adenomas of the colon, and in one tumour of each kind mutated c-K-*ras* and N-*ras* were both identified; they did not, however, find mutated *ras* genes in adjacent normal colonic mucosa. The state of progression of the tumour, as judged by its invasiveness and the presence of metastases, did not appear to correlate with the incidence of mutated *ras* genes, though it did seem to be related to the position in codon 12 at which the mutation occurred.

These findings are persuasive, but not conclusive. Activated proto-oncogenes are not consistently associated with particular tumours—in SCLC, for example, there is evidence of activation of a *ras* proto-oncogene by point mutation in a few, but only a few, patients; and it is still true, as Duesberg said in 1985, that no one has yet isolated from the same tumour an activated proto-oncogene and a complementary gene which can act synergistically to bring about transformation.

Moreover, as already mentioned (Section 2.4.1, p. 12), it can be argued that what Duesberg calls *cancer genes*, but which, to avoid getting lost in a semantic jungle, we shall continue to call *oncogenes*, are not activated normal cellular genes but arise from normal cellular genes by truncation and recombination. On this view, 'no cellular gene is a structural or functional homologue of a viral *onc* gene, but the viral *onc* genes appear to be models of how cancer genes may arise from normal cellular genes by rare truncation and recombination' (Duesberg 1987). It is difficult, however, to see how, if Duesberg's thesis is generally true, an oncogene could arise as the result of a chromosomal translocation that brings a proto-oncogene under the influence of promoter-enhancer sequences of some other gene, as happens, for example, when c-*abl* is translocated from its normal position on chromosome 9 to a position on chromosome 22 (see Section 2.4.1, p. 13).

The development of techniques for producing transgenic mice by introducing DNA into fertilized eggs or early embryos has made it possible to study the effect *in vivo* of oncogenes expressed in cells of a particular kind, or at a particular stage of differentiation, by combining the oncogene with tissue-specific promoter-enhancer elements from other genes. In studies of this kind with *ras* and *myc* oncogenes (rev. Groner *et al*. 1987), the incidence of tumours in the tissues in which oncogene expression was elicited was usually, but not always, greater than in non-transgenic mice of the same strain, but there was often a delay of many months between expression of the oncogene and the appearance of a tumour. This delay, together with the fact that the tumours were monoclonal (Chapter 3) despite the broad polyclonal activation of the transgene (Klein 1987), suggests that activation of additional oncogenes or loss of suppressor genes may be needed to produce a tumour in this system.

It will be of great interest to see how this work develops. The results obtained so far strengthen still further the case for the view that activated proto-oncogenes are important in carcinogenesis, but just how important they are is still an open question.

2.4.2 Tumour suppressor genes

Many years ago, in a remarkable paper, Comings (1973) put forward a general theory of carcinogenesis, according to which this process is due to the action of multiple transforming genes that exist in all cells and are active during embryogenesis but are later normally suppressed by diploid pairs of regulatory genes. Spontaneous tumours, and tumours induced by chemicals or radiation, he postulated, arise as the result of a double mutation of any set of regulatory genes which releases the suppression of the corresponding transforming gene. Autosomal dominant hereditary tumours are the result of germ line inheritance of one inactive regulatory gene and subsequent mutation of the other regulatory gene, as postulated by Knudson (1971, 1973).

The application of molecular techniques has confirmed Comings' central idea of the existence of genes that suppress carcinogenesis, though these do not necessarily act by suppressing oncogenes. This is why the neutral term *tumour suppressor gene* is preferable to the term *anti-oncogene*. Another name that has been proposed and is appropriately descriptive, but seems unnecessary, is *emerogene*, derived from the Greek verb meaning to tame or domesticate.

Decisive evidence of the existence of tumour suppressor genes has been obtained in two main ways:

(1) by demonstrating that both alleles of a particular gene have been lost from

the cells of certain tumours; the best known, but not the only, examples being retinoblastoma and Wilms' tumour (Section 2.4.2, pp. 20–21);

(2) experimentally, by analysing the effect on tumorigenicity of cell hybridization, and microcell transfer of a single chromosome (Section 2.4.2, pp. 22–23).

Supporting evidence (rev. Klein 1987) has come from the study of *in vitro*-transformed cells that revert to a non-tumorigenic phenotype (Bassin and Noda 1987), and the induction of terminal differentiation in neoplastic cells (Section 2.6); the demonstration of regulatory sequences in the immediate vicinity of certain oncogenes (Vande Woude *et al.* 1987); and the discovery, discussed in Sections 2.6.1 (p. 26) and 2.6.2, that diffusible products released by surrounding normal cells can, under certain conditions, gain entry to transformed or pre-neoplastic cells via gap junctions and inhibit their growth. There is also evidence which suggests that the same gene may act in either a transforming or a suppressor capacity, depending on the target cell. Thus, sarcoma viruses carrying *ras* oncogenes have been shown to induce differentiation in a rat phaeochromocytoma cell line (Noda *et al.* 1985).

Deletion of suppressor genes in human tumours

The study of inherited predisposition to various tumours has provided important leads to our understanding of the role of suppressor genes in cancer, and knowledge in this field is growing rapidly.

Retinoblastoma. Retinoblastoma is a malignant tumour of children that occurs in two forms: familial and sporadic. It was postulated by Knudson (Knudson 1971, 1973; Hethcote and Knudson 1978) that in either case development of a tumour depends on two genetic events; in the familial form one occurs in the germ line and the other later in a retinoblast, whereas in the sporadic form both occur as somatic mutations in the same retinoblast.

It now seems clear that this hypothesis is correct, and that the essential lesion is loss of both copies of a gene called *Rb* that is located in the q14 region of chromosome 13.

In some familial cases there is a deletion at q14 on one chromosome 13 in all the patient's cells, and in some sporadic cases there is a similar deletion, or total loss of one chromosome 13, in the tumour cells but not in the patient's normal cells. Evidence of loss of *Rb* from a chromosome that does not show an obvious morphological deletion has been obtained in various ways (see Murphree and Benedict 1984; Evans 1986; Monier 1987). For our purposes a brief summary of this work will suffice.

1. The *Rb* locus is closely linked to a locus on chromosome 13 that codes

for esterase D (ESD). In the cells of patients with familial retinoblastoma ESD levels were 50 per cent of normal, whether or not there was a demonstrable 13q14 deletion. Moreover, in one patient studied, the tumour cells showed no ESD and had only one chromosome 13; i.e. the tumour cells had lost the normal chromosome 13 and were monosomic for a normal-looking but defective chromosome 13 that had no *ESD* locus, and an absent or mutant *Rb* locus (Sparkes *et al.* 1983).

2. It has been shown in other patients, by using DNA probes which reveal polymorphic restriction endonuclease recognition sequences, that the tumour cells were homozygous for either the whole defective chromosome 13 or for the mutant segment, and that this was attributable either to loss of the normal 13 by non-dysjunction and duplication of the abnormal 13, or to reciprocal chromatid translocation (Cavanee *et al.* 1983).

3. It has been shown that the transcription product of a gene located in a segment of chromosome 13 that appears to include the *Rb* locus is absent or altered in retinoblastomas but present in human fetal retina, placenta, and various tumours other than retinoblastoma and osteosarcoma (Friend *et al.* 1986; Lee *et al.* 1987). This gene has now been cloned, and evidence that it is indeed the retinoblastoma gene has come from the discovery in the cells of two retinoblastomas of deletions that begin and end within the confines of the cloned gene (Weinberg 1988). If this conclusion is correct, insertion of the cloned gene into tumour cells lacking the *Rb* gene should cause them to revert to a normal phenotype. This prediction is currently being tested.

The role of the *Rb* gene does not seem to be limited to its effect on retinal cells, because in patients with familial retinoblastoma who survive long enough there is a high incidence of osteosarcoma and, to a lesser extent, of other tumours (see Murphree and Benedict 1984).

Wilms' tumour. Like retinoblastoma, Wilms' tumour also occurs in two forms: a familial form, which is associated with aniridia, and a sporadic form which is not. A suppressor gene, located in the p13 region of chromosome 11, plays a role similar to that of *Rb* in retinoblastoma (see Koufos *et al.* 1984). At the time of writing, this gene had not been isolated, but several groups seemed to be within sight of this goal.

Other tumours. As mentioned in Section 2.4.1, pp. 17–18, in most small cell carcinomas of the lung there is a characteristic deletion in the short arm of chromosome 3 that is not present in the patient's normal cells. Although this is seen in only one chromosome, analysis by restriction fragment length polymorphism has shown that there is also a deletion in the other chromosome 13 that includes a tumour suppressor gene (Naylor et al. 1987).

Following the demonstration of a deletion in the q region of chromosome 5 in a patient with adenopolyposis of the colon (Bodmer et al. 1987)—in which there is a high risk of carcinoma of the colon—it has been shown that loss of both copies of a gene in this region may be a critical step in the progression of quite a high proportion of colorectal carcinomas (Solomon et al. 1987).

There is good evidence that deletions on chromosomes 22, 17, and 10 may play a similar role in the development of acoustic neuroma (Seiziger et al. 1986; Rouleau et al. 1987), generalized neurofibromatosis (Barker et al. 1987; Seizinger et al. 1987), and multiple endocrine neoplasia (Mathew et al. 1987; Simpson et al. 1987), respectively. There is also evidence that a deletion on chromosome 13 may be associated with some cases of osteosarcoma without retinoblastoma (Dryja et al. 1986), and that carcinomas of the bladder and breast may, on occasion, be associated with loss of suppressor genes on chromosome 11.

With the suppressor genes that have so far been discovered in human tumours, a single copy of the gene suffices to suppress carcinogenesis; it therefore seems reasonable to describe them as dominant suppressor genes. Many authors, however, still prefer to speak of mutations recessive to wild-type alleles that predispose to tumour growth. According to the first view, a person who is hereditarily Rb/rb or Rb/– (where Rb denotes the wild-type normal Rb gene, rb denotes a mutated inactive form, and – denotes absence of the gene) has lost one copy of a tumour suppressor gene; on the second view, he is said to carry the recessive retinoblastoma gene. To avoid confusion it seems advisable to avoid using the terms dominant and recessive in this context.

Evidence from cell hybridization and chromosome transfer

As we have seen (Section 1.2.5), hybrids of tumour and normal cells are usually not tumorigenic if they retain all, or nearly all, of the complete set of chromosomes of each cell, but may regain tumorigenicity as the result of selective chromosome loss. A deeper analysis of this phenomenon has been made possible by advances in molecular cytogenetics (see Sager 1985) and the development of a technique by which a single chromosome can be introduced into a recipient cell (Fournier and Ruddle 1977).

The following observations by Craig and Sager (1985), and by Stanbridge's group (rev. Stanbridge 1987), are of special significance:

1. Hybrids of normal rodent fibroblasts and CHEF cells transformed with a mutated Ha-*ras* oncogene (Craig and Sager 1985), or of normal human fibroblasts and cells of a human bladder carcinoma line that express the same Ha-*ras* oncogene (Geiser et al. 1986), are much less tumorigenic than the original transformed or tumour cells, despite the fact that they express the *ras* gene product at much the same level.

2. Transfer into either HeLa cells or tumorigenic segregants of HeLa cell–human fibroblast hybrids of a single fibroblast chromosome, t(X;11), that contains about 95 per cent of human chromosome 11 plus the q26-qterm part of the X chromosome in which the HPRT gene is located, completely suppressed tumorigenicity (Saxon et al. 1986). This chromosome, which had appeared as the result of a translocation in a fibroblast cell line, was used in preference to a normal chromosome 11 because the authors were then able to select cells that had lost the introduced chromosome by culture in medium containing 6-thioguanine, and to show that these cells were tumorigenic. Control experiments showed that transfer of an X or any other chromosome instead of t(X;11) did not suppress tumorigenicity. It is of interest that the HeLa cell–fibroblast hybrids were tumorigenic despite the fact that they contained one chromosome 11. To explain this, Saxon et al. postulate that other fibroblast-derived chromosomes in some way modulate the level of expression of the chromosome 11-specific suppressor activity.

In later experiments (Weissman et al. 1987), it was shown that introduction of a single chromosome t(X;11) rendered the cells of a Wilms' tumour line non-tumorigenic, as judged by their inability to grow in athymic mice, though it did not alter their behaviour in tissue culture.

George Klein (1987) has suggested that the study of suppressor genes, though experimentally more difficult than 'the pursuit of the oncogene', may turn out to be even more rewarding. Few, if any, people familiar with the field would dissent from this cautious prediction by a leading tumour biologist. It seems right to add that work on suppressor genes seems likely to have important therapeutic implications that will not be long delayed.

2.4.3 Modulator genes

Modulator genes (see Klein and Klein 1985) constitute a large and heterogeneous group of genes which includes genes of the major histocompatibility complex (MHC). They do not cause transformation but may affect the local invasiveness and metastatic capacity of tumours (Chapter 5), and their immunogenicity and resistance to immunological rejection.

2.5 Environmental factors affecting transformation

The host may play a crucial role in initiating transformation *in vivo* by converting substances that are not themselves carcinogenic into active carcinogens through the action of microsomal enzymes in the liver and elsewhere, and by the action of hormones and other growth factors. Moreover, as transformation proceeds, the emerging population of partially and fully transformed cells is dependent on the host for its nutritional requirements and continues to be exposed to the influence of hormones and other factors of host origin.

The members of this population fall into the category of what have been called *asocial cells* (Stoker 1972) or *ideosomatic predators* (Melicow 1982), and it would be surprising if the metazoa had not evolved homeostatic mechanisms to limit the development and multiplication of such cells. That they have done so is suggested by the unexpectedly low incidence of spontaneous tumours in relation to the frequency of mutation. The average human adult has about 3×10^{13} cells and, allowing for the huge number of cells lost from the skin and intestinal epithelium, it has been estimated that between 10^{15} and 10^{16} cell divisions occur during the average human life span (Cairns 1975); assuming a frequency of mutation of between 10^{-5} and 10^{-6} per locus per cell generation, one would therefore expect at least 10^{10} mutations. There is no obvious reason why a high proportion of these should be potentially carcinogenic but, unless this proportion is extremely small or the number of mutations required for carcinogenesis to occur is very large, one would expect the incidence of cancer to be considerably greater than it is if no defence mechanisms existed (Woodruff 1980).

Studies of experimental carcinogenesis and DNA repair suggest ways in which the emergence of cancer cells may be prevented, and how these mechanisms may fail.

1. A chemical carcinogen may be rendered innocuous by cells in the liver and elsewhere.

2. At a cellular level there are various DNA repair mechanisms which normally operate to prevent the emergence of cells bearing potentially dangerous mutations. These may be defective, as is the case, for example, in patients with the rare recessive hereditary disease *xeroderma pigmentosum*, who are unable to repair DNA damage caused by ultraviolet irradiation and, in consequence, have a high incidence of skin cancer.

3. There may be selective shedding of mutant cells from epithelial surfaces, as suggested by Cairns (1975). According to this hypothesis, for which there does not seem to be any evidence, when an 'immortal' stem cell divides to form another stem cell and a 'mortal' cell which is destined to be shed from the skin or some other epithelial surface, and a copying error has arisen during DNA replication in one strand, the abnormal strand goes selectively to the 'mortal' cell.

4. Anatomical constraints limit the ability of stem cells to escape into areas populated by other stem cells, so that competition between a potentially vigorous mutant and normal cells is limited. Thus, to explain the relatively low incidence of cancer of the small bowel despite the rapid cellular turnover, Peto (1977) has suggested that we may have evolved 'stem cell hierarchies and territorial

imperatives, based on crypts and villi, that allow this [rapid turnover] without much risk of cancer, whereas perhaps no such mechanisms have been evolved to protect the bronchi because apes and coelocanths did not smoke cigarettes'.

5. Partly and fully transformed cells may be eliminated by immunological or para-immunological mechanisms (Woodruff 1980).

2.6 Is transformation reversible?

There is convincing evidence that malignant stem cells can give rise to non-malignant end-state cells, and also to non-malignant stem cells.

The first of these possibilities was suggested to the writer some 40 years ago by a distinguished radiotherapist, Ralston Paterson, who postulated that it played a role in the healing of carcinomatous ulcers, in particular ulcerating squamous cell carcinomas of the skin, in response to radiotherapy. This does not seem to have been investigated in patients, but it has been shown, by administrating ^{3}H-thymidine to experimental animals bearing squamous cell carcinomas and following the label by autoradiography, that keratinocytes in the epithelial pearls of these tumours differentiate from malignant cells (Pierce and Wallace 1971).

The second proposition is supported by observations *in vitro* and *in vivo*.

2.6.1 *In vitro* evidence of reversion

As we have seen (Section 2.4.2, pp. 22–23), a tumorigenic cell may be rendered non-tumorigenic by the insertion of a chromosome containing an appropriate tumour suppressor gene, so transformation can be reversed *in vitro* by genetic manipulation.

Transformation can also be reversed *in vitro* by *differentiation inducers* of various kinds.

Myeloid cells

The work of Metcalfe and of Sachs (rev. Metcalfe 1987; Sachs 1987*a,b*) has led to the identification of four growth-stimulating proteins that promote the viability and multiplication of myeloid cells, and to the further demonstration that myeloid cells stimulated by these proteins, and some non-myeloid cells, produce other proteins that cause myeloid cells to differentiate.

Myeloid leukaemic cells, according to Sachs, comprise:

(1) cells that need little or no growth stimulator;

(2) cells that produce their own growth stimulator;

(3) cells that do not produce a differentiation-inducer in response to a growth stimulator;

(4) differentiation-defective cells that cannot be made to differentiate fully, and sometimes not at all, by normal differentiation-inducing proteins. These are termed D^- cells; the others are all D^+

Many agents other than normal differentiation-inducing proteins, including steroids, hormones, cytotoxic drugs in low dosage, insulin, retinoic acid, and X-irradiation, can cause D^+ cells to differentiate, and some can cause D^- cells to do so by alternate differentiation pathways. Different substances may affect different parts of the differentiation programme, so that combinations of substances may produce by complementation the balance of expression of oncogenes and suppressor genes needed for the cells to revert to a non-malignant, though not necessarily normal, phenotype.

Other cells

Much less is known about the reversibility of transformation in other types of cell.

It has been reported that reversion to a non-transformed phenotype can occur *in vitro* with fibroblasts transformed with polyoma virus or chemical carcinogens (Sachs 1974) and with sarcoma cells (Sachs 1984) if they are cultured at low density, though the poloma-transformed cells retain the viral genome. Analysis of revertants of cells transformed with v-*ras* that still expressed the protein of the transforming oncogene at the original level points to the conclusion that reversion in this case was due to the action of a suppressor gene (Bassin and Noda 1987).

Further evidence comes from the work of Yamasaki and his colleagues. As already noted (Section 1.2.2), they have reported that, when fibroblasts of the NIH 3T3 line are transformed *in vitro* by a chemical carcinogen or an activated oncogene, the transformed cells communicate with each other via gap junctions but do not communicate with untransformed cells (Yamasaki 1986; Yamasaki *et al.* 1987). They have gone on to show that communication between transformed and normal cells may be established by adding cAMP, dexamethasone or retinoic acid to the culture medium, and that when this happens some of the foci of transformed cells disappear (Yamasaki and Katoh 1988).

Evidence relating to teratocarcinoma, neuroblastoma and other tumour cells may be found in a review by Pierce and Speers (1988).

2.6.2 *In vivo* evidence of reversion

It has been established by the work of Mintz and others that murine teratocarcinoma cells transferred to mouse blastocysts can give rise to the whole gamut of differentiated cells, including mature spermatozoa, in the chimaeric animals that develop as a result of this procedure (Mintz and Illmensee 1975; Papaioannou et al. 1975; Illmensee and Mintz 1976; Stewart and Mintz 1981).

Sachs and his colleagues have studied the fate of murine myeloid leukaemic cells bearing an alloenzyme marker injected into mid-term mouse embryos. Many of the injected mice died of leukaemia, but differentiated myeloid cells of donor origin were identified in those which survived (Gootwine et al. 1982; Webb et al. 1984).

Further evidence of reversion *in vivo* comes from the study of tumour recurrences and metastases.

Cushing and Wolback (1927) reported that after an unsuccessful attempt to remove a paravertebral neuroblastoma from an infant, the tumour was found many years later, in the course of an appendicectomy, to exhibit the morphological features of a benign ganglioneuroma. Several similar cases have been reported subsequently. It has also been reported that metastases from malignant testicular germ-line tumours may assume the appearance of benign teratomas (Karpas and Jawahiri 1964; Smithers 1969; Carr et al. 1981). This phenomenon seems to have become more common in recent years, and Carr et al. suggest that this may be a consequence of the widespread use of chemotherapy.

2.6.3 Therapeutic implications

The work we have been considering raises the question of whether agents that can promote differentiation could be of therapeutic value (Sachs 1987*b*; Sartorelli et al. 1987).

Some encouraging results have been reported in animal experiments with retinoic acid (Speers 1982), and in human patients with myeloid leukaemia treated with low doses of cytosine arabinoside—a substance that can induce myeloid leukaemic cells to differentiate *in vitro*.

Differentiation-inducing agents would seem to offer the possibility of neutralizing neoplastic cells at relatively little risk to normal cells. It seems unlikely, however, that all the neoplastic cells in a patient will prove responsive, and the risk that the treatment will select resistant cells that behave in a very aggressive way would seem to be considerable.

3
Clonal composition of tumours

3.1 Definitions

The members of a clone are, by definition, all descendants of one cell. We can choose any cell we like as the founder of a clone (Micklem 1986), but need to state explicitly what category of cell we have chosen.

What is usually implied, though rarely stated explicitly, in discussing the clonal composition of tumours, is that the founder cell of a neoplastic clone is an initially normal cell that takes one or more steps on the road to transformation before it divides, and some of whose descendants become fully transformed. This corresponds to clonality as determined with X-linked markers (Section 3.2.1, pp. 29–31), and would be the only possibility worth considering if carcinogenesis were a one-step process. In so far as carcinogenesis is multistep, however, we could start instead with a fully transformed cell, or, if a stable marker appears for the first time during the course of transformation, with a partly transformed cell.

Writers on tumour clonality often do not state clearly, and sometimes seem unaware, that they have chosen one of these alternative starting points; this has been the cause of much confusion, and probably accounts for the common, but unsupported and probably false, assertion that most tumours originate from a single *transformed cell*. In the interests of clarity we shall, following Woodruff (1988), refer to clones as N-clones, P-clones, or T-clones, according to whether the chosen founder cell was a normal cell, a partly transformed cell, or a fully transformed cell. A tumour whose neoplastic cells all belong to a single N, P, or T clone will be labelled N, P, or T *monoclonal*, respectively; if the population consists of more than one clone the tumour will be labelled *pleoclonal*. The number of different clones represented will be termed the *clonality* of the tumour.

Another question to be considered is whether, as Deamant *et al*. (1986) have maintained, a valid distinction can be drawn between a *single* tumour and a *conglomerate* tumour formed by the coalescence of two or more single tumours. Any such distinction would have to be based on the spatial distribution of recognizably different clonal populations, but even if the necessary markers were available this would not dispose of the problem. It may happen that two or more tumours arise at widely separated sites and subsequently coalesce to form a single mass, and in this event the term *conglomerate tumour* would certainly be appropriate, but what label should we use if two adjacent cells become transformed and proliferate to form populations that begin to intermingle after a few cell generations? If the resulting tumour is classed as conglomerate there can be no such thing as a pleoclonal single tumour; if it is labelled single then where are we going to draw the line between single and conglomerate tumours?

3.2 Evidence of monoclonality

3.2.1 Clonal markers and their limitations

The markers used to study the clonality of human and/or animal tumours may be classified as follows:

1. X-lined markers:
    (a) Alloenzymes [glucose-6-phosphate dehydrogenase (G-6-PD) in humans and phosphoglycerate kinase 1 (PGK-1) in mice];
    (b) Restriction fragment length polymorphisms (RFLP).
2. Karyotypic and other chromosomal markers.
3. Immunoglobulins (Ig); Ig gene and T-cell receptor gene rearrangements.
4. Viral integration sites.
5. Tumour-associated antigens.
6. Markers in natural or experimental chimeras.

We shall discuss the principles underlying the use of these markers for studying tumour clonality and the interpretation of the results. For more detailed information the reader is referred to reviews by Fialkow (1972, 1976) and Woodruff (1988).

X-linked markers

The use of X-linked markers depends on the fact (see Lyon 1974) that in the somatic cells of mammals only one X-chromosome is active, no matter how many are present. The differentiation of a particular X chromosome as active or inactive begins early in embryogenesis, and once the choice is made it is maintained through all subsequent cell divisions. Except in marsupials, and in the extraembryonic tissues of rodents, the choice of whether the paternally derived X (X^P) or the maternally derived X (X^M) is inactivated appears to be, mathematically speaking, random, though the probability of being inactivated is not necessarily the same for X^P and X^M, notably in mice. When the probabilities are unequal, geneticists, it may be noted, are apt to refer to the phenomenon as *non-random inactivation*.

If a person or animal is heterozygous for two alleles of an X-linked marker, i.e. if X^P codes for one form of a marker and X^M for the other, then a tumour cell population that is N-monoclonal will reveal only one form of the marker, namely the form carried by the normal cell from which the tumour originated. A pleoclonal population may reveal one or both forms, and the greater the clonality of the tumour, the greater the probability that it will reveal both. As we shall see later (Section 3.4 and Fig. 3), assays with X-linked markers do not, as a rule, tell us anything about T-clonality. The only circumstances in which such assays could show that a tumour had arisen from a single transformed cell would be if the whole process of transformation, once begun, was completed before the cell divided.

The use of X-linked markers is restricted because the hosts must be female and the alleles must be sufficiently common to yield enough heterozygotes for study. In humans, a lot of work has been done using the alloenzymes of G-6-PD, but the number of patients available for study is limited by the restricted ethnic and geographical distribution of the A form of the enzyme. The situation has been greatly improved by the discovery of RFLPs which can currently be used to study clonality in about 25 per cent of female patients, and it should soon be possible to undertake such investigations in virtually all women with tumours. In mice, sex is the only limitation, because inbred strains homozygous for PGK-1 A and PGK-1 B are now available, and all that is needed is to raise F1 hybrids from parents expressing different forms of the enzyme.

To decide what weights to attach to the evidence of N-clonality provided by X-linked assays we must consider what factors other than clonality may influence the results.

The main reason for observing a double phenotype with a truly monoclonal tumour is admixture of normal stromal cells with the tumour cells. It is difficult to eliminate the possibility of error due to this cause completely, though there are various ways in which it can be reduced. Another, perhaps remote, possibility is that, contrary to the general rule enunciated by Lyon, two or more X-chromosomes remain active in some or all of the tumour cells.

The converse situation, in which a pleoclonal tumour appears to express only a single phenotype, may be due to any of the following causes:

1. A small subpopulation escapes detection because the method of assay is not sufficiently sensitive.

2. Cells expressing X^P or X^M are clustered in patches and the tumour developed from two or more normal cells in the same patch.

3. Selection may favour transformation and survival of cells of a particular phenotype.

4. X-chromosome loss or anomalous gene expression may occur in the neoplastic cells.

Karyotypic and other chromosomal markers

Chromosomal abnormalities in cancer may be constitutional, like those that occur in the inherited forms of retinoblastoma and Wilms' tumour (Section 2.4.2, pp. 20–21), or acquired. Acquired abnormalities include microscopically recognizable translocations (see Section 2.4.1, p. 17), of which the common 9/22 translocation responsible for the Philadelphia chromosome (Ph^1), seen in almost all the mitoses in the bone marrow of over 85 per cent of patients with chronic myeloid leukaemia, is the best-known example, and allelic loss recognizable at the molecular level in many patients with retinoblastoma, Wilms' tumour and osteosarcoma.

Patients with constitutional abnormalities are congenitally heterozygous for a specific chromosomal deletion in all their normal somatic cells, and the presence of the deletion in tumour cells gives no information whatever about tumour clonality.

The acquired abnormalities do not provide any evidence for or against N-monoclonality; they do, however, provide evidence about P-clonality and, especially if used in conjunction with an X-linked marker, about changes in clonal composition during carcinogenesis. At present, their usefulness is greatly limited by uncertainty about when they first appear.

Ig; Ig gene and T-cell receptor gene rearrangements

Membrane-bound Ig on a normal human B lymphocyte is restricted to one Ig class (and, if this is IgG, to one subclass), and one Gm allotype; Ig on normal murine B lymphocytes is similarly restricted to one Ig class and one light chain. Surface Ig, and Ig gene rearrangements, can therefore provide good evidence concerning the N-clonality of tumours whose cells secrete Ig or possess cell surface Ig.

As with other markers, factors other than tumour clonality may influence the results. On the one hand, one or more clones might not produce detectable amounts of Ig, in which case a pleoclonal tumour might be mistakenly diagnosed as monoclonal. Conversely, the presence of two or more Ig molecules is not necessarily inconsistent with N-monoclonality because Ig-gene rearrangement or class switching may have occurred during or after transformation.

The T-cell antigen receptor has structural homologies with Ig domains. It has a V region and a C region, but it is encoded by genes which are completely separate from the Ig genes. The receptor possesses two disulphide-linked polypeptide chains, alpha and beta, coded for by DNA segments T_α and T_β, and the gene arrangement is very similar to that of the Ig genes (Roitt et al. 1985). In the course of T-cell receptor gene rearrangement, part of the T_β constant region is deleted, and this deletion can serve as a marker for cells of T-lineage.

In a pleoclonal population of normal T-cells, individual beta variable region recombinants are not detectable by current techniques, but a monoclonal population of T-cells may yield a detectable band corresponding to the T_β variable region. Raskind and Fialkow (1987) have made use of this observation to study the clonality of a variety of T-cell neoplasms, and their findings point to the conclusion that many, though not all, of these were, in our terminology, N-monoclonal.

Viral integration sites

Tumours induced by naturally occurring retroviruses frequently develop from cells in which the provirus has integrated adjacent to, and has consequently activated, a particular cellular oncogene. This, however, is not a universal rule. The human T-cell lymphoma virus (HTLV), for example, which appears to be involved in the development of various neoplasms of T-cell origin (see Broder 1984), has, in addition to the three structural genes, a viral gene region, X, whose product may activate cellular genes and cause transformation irrespective of where the provirus happens to be integrated (Yoshida *et al.* 1984). In such cases the site of integration can constitute a marker of N-clonality or P-clonality, depending on whether the integration occurs in a normal cell or a cell that has already taken one or more steps on the way to transformation.

Raskind and Fialkow (1987) used Southern blotting (see, for example, Franks and Teich 1986) to determine the site of integration of HTLV in various 'T-cell proliferative syndromes', and concluded that most of these were monoclonal or oligoclonal, but did not completely rule out the possibility that they might 'initially be polyclonal with subsequent overgrowth by a single dominant clone'.

The same technique might provide valuable evidence of the clonality of various neoplasms in which DNA viruses may play an aetiological role, including human nasopharyngeal, hepatocellular and cervical carcinomas.

Tumour-associated antigens

As mentioned earlier (Section 1.1), some experimental tumours possess unique tumour-associated transplantation antigens (TATA).

Many studies have been concerned with chemically induced fibrosarcomas (rev. Woodruff 1980). As was first reported by Prehn and Main (1957), there is often, though not invariably, no demonstrable cross-reactivity between the TATA of different sarcomas induced with the same carcinogen, including sarcomas induced at different sites in the same host (Globerson and Feldman 1964; Rosenau and Morton 1966). It has been shown further that tumour sublines established *in vivo* by transplanting tissues from opposite poles of the same tumour may possess different TATA (Prehn 1970; Pimm *et al.* 1980).

More recently, in studies of fibrosarcomas induced with methylcholanthrene in female mice heterozygous for the X-linked alloenzymes of PGK-1 (see Section 3.2.1, p. 30), which were certainly pleoclonal and probably N-biclonal, subpopulations of the same tumour obtained by cloning *in vitro* appeared to possess identical TATA if, but only if, they possessed the same alloenzyme marker, and were therefore almost certainly descendants of the same normal cell (Woodruff *et al.* 1984). It would seem, therefore, that the neoplastic descendants of a single normal progenitor may share a marker of a kind that is not possessed by normal cells, or by neoplastic cells derived from a different normal progenitor.

It would follow, if these results are confirmed, that in the development of chemically induced sarcomas, cells become committed to develop a particular TATA or set of TATA when they take the first step on the road to transformation and before subsequent division. The factors responsible for this commitment might conceivably include the state the cell had reached in the cell cycle when it first encountered the carcinogen, and parameters relating to the local environment, for example the concentration of carcinogen, which might differ significantly over quite small distances. In this regard, it is important to remember that many so-called chemical carcinogens are, strictly speaking, *procarcinogens*; i.e. they possess little or no carcinogenic activity themselves but become converted in the body by the action of microsomal enzymes to *proximate* and *ultimate carcinogens*, which bind covalently to cellular DNA, RNA or protein, and, if the cell is not killed, initiate the process of carcinogenesis (rev. Woodruff 1980, p. 19, *et seq.*).

Further work is urgently needed, firstly to see whether the results reported can be confirmed, and secondly to try to elucidate the molecular basis for the remarkable polymorphism displayed by the TATA of chemically induced murine sarcomas. It might, at the same time, be rewarding to perform similar experiments with other kinds of tumour that possess TATA.

Markers in natural and experimental chimeras

Natural human chimeras may arise in various ways (McLaren 1975), but are too rare to be of much practical importance for the study of clonality.

Tumours induced in experimentally produced animal chimeras have been studied by Iannaccone and others (see Iannaccone *et al.* 1987). In interpreting the results it is important to bear in mind the possibility that the two cell subpopulations existing in the chimera may differ in their liability to carcinogenesis. The risk of error seems likely to be greater with interspecific than with intraspecific chimeras.

3.2.2 Conclusions

Human tumours

Only two human tumours, congenital neurofibroma and the rare trichoepithelioma, have been shown to be regularly N-pleoclonal. Many others, both benign and malignant, including fibromyomas (in which monoclonality was first demonstrated), warts, Burkitt's and other lymphomas, chronic myeloid leukaemia, Wilms' tumour, and carcinomas of the cervix, lip and mouth, have, with occasional exceptions (including probably some carcinomas of the colon), been reported to be monoclonal (rev. Fialkow 1972, 1976; Raskind and Fialkow 1987).

The conclusion that human tumours are, as a general rule, monoclonal is, however, subject to two provisos:

1. Some tumours may be pleoclonal at an early stage in their development. This seems to be the case when colorectal cancer develops in a patient with familial adenopolyposis of the colon, since the precancerous polyps have been reported to be pleoclonal (Hsu et al. 1983). On the other hand, carcinoma of the cervix has been reported to be monoclonal even at the pre-invasive stage.

2. As pointed out in Section 3.2.1, the existence of clones too sparse to be detected by the methods used for the assays cannot be excluded. It would be unwise to dismiss this as no more than a remote theoretical possibility. It provides a plausible, though not the only conceivable, explanation of the observation (Fialkow 1972) that lymphoblastoid cell lines from human B-cell lymphomas may differ from the cells of the original lymphoma in respect of their Ig, and also, if the tumour arose in a G-6-PD heterozygote, in alloenzyme phenotype; and of a similar phenomenon that has been observed with animal tumours (Section 3.3.3).

Animal tumours

Animal tumours offer great scope for studying clonality. With induced tumours the factors that influence clonality can be investigated under controlled conditions, and with both naturally occurring and experimentally induced tumours the tumorigenicity of cloned subpopulations can be tested by transplantation (Woodruff et al. 1982).

This potentially valuable material has not been studied as thoroughly as it deserves, and the results obtained in different laboratories have sometimes been surprisingly discordant.

Papillomas and basal cell carcinomas in chimeric mice have been reported to be nearly always monoclonal by Iannaccone et al. (1987), whereas Reddy and Fialkow (1983) found that about 14 per cent of papillomas in PGK-1 heterozygotes were pleoclonal. The difference may be due in part to the use of different markers, but this does not explain why studies of fibrosarcomas in PGK-1 heterozygotes have led some authors to conclude that these tumours are often pleoclonal and others to conclude that they are nearly always monoclonal. Comparison of the experimental protocols used in different laboratories suggests that the dose of carcinogen may be of critical importance, and that the tumours are more likely to be monoclonal when the dose is relatively small.

The B-cell lymphomas that develop in transgenic mice bearing the c-*myc* oncogene coupled to the lymphoid-specific Ig heavy chain enhancer are of particular interest. The fully developed tumours have been reported to be monoclonal on the basis of IgH-locus rearrangement patterns (Adams *et al.* 1985), but the development of tumours is preceded by the appearance of an expanded polyclonal population of B-cell precursors (Langdon *et al.* 1986).

3.3 Pleoclonal tumours

With pleoclonal tumours three important questions arise: how many clones are there, how are they distributed spatially, and does the clonal composition remain constant?

3.3.1 How many clones?

The problem of determining the number of clones in a pleoclonal tumour could be solved only if each clone carried a unique, stable, detectable marker.

Tumours that secrete Ig may fall in this category, though, as we have seen (Section 3.2.1, p. 31), it is not certain that every clone will produce detectable amounts of Ig, and it is just possible that a single N-clone may produce more than one Ig because Ig gene rearrangement or class switching may have occurred during or after transformation. With this proviso, studies of the cell surface Ig and Ig gene configuration of human B-cell lymphomas (Sklar *et al.* 1984) point to the conclusion that some of these tumours are biclonal.

Tumours that possess TATA may also conceivably fall in this category, though this is still largely a matter of speculation.

With these possible exceptions the problem is at present insoluble, but the outlook is not quite so bleak if, instead of asking how many clones are represented in a particular tumour, we set out to determine, for a set of similar tumours, the proportions $(C_1, C_2, C_3, \ldots, C_k, \ldots)$ of tumours that are monoclonal, biclonal, triclonal, etc.

This question has been attacked in respect of methylcholanthrene-induced murine fibrosarcomas by raising tumours in female mice heterozygous for the alloenzymes of PGK-1 and determining the proportions $(p_A, p_B$ and $p_{AB})$ of tumours that express only PGK-1 A, only PGK-1 B or both PGK-1 A and PGK-1 B (Woodruff *et al.* 1986*a*). A summary of the data is shown in Table 1, together with the expected values of p_A, p_B and p_{AB}, calculated on the assumption that (1) the proportion (r) of normal cells expressing PGK-1 A in the population at risk of transformation equals the mean proportion (0.72) of A alloenzyme found in the blood of the F1 hybrid mice in which the tumours were raised; and (2) the risk of transformation, and the probability that a transformed cell

Table 1. Observed and expected numbers of murine fibrosarcomas expressing one or both alloenzymes of PGK-1*

Alloenzymes expressed	Observed proportion of tumours	Expected proportions of tumours for clonality shown at head of column					
		1	2	3	4	5	10
A	29/48 = 0.60	0.72	0.52	0.37	0.27	0.19	0.04
B	2/48 = 0.04	0.28	0.08	0.02	0.01	0.00	0.00
A and B	17/48 = 0.36	0.00	0.40	0.61	0.72	0.81	0.96

*From Woodruff et al. 1986a.

will give rise to a detectable clone, are the same for A-cells and B-cells. The expected values are very sensitive to changes in the value assigned to r, and the estimate of r is subject to considerable error, but inspection of the table suggests that, unless the value chosen is wildly out, many of the tumours had just two clones. Algebraic analysis supports this conclusion. This is given in full elsewhere (Woodruff et al. 1986a; Woodruff 1988). In essence, since the data generate only three equations from which to calculate $C_1, C_2, C_3, \ldots C_k$, no general solution is possible unless k is not greater than 3. It would be unwarranted to assume this *a priori*, but it can be shown by elementary algebra that C_3, C_4, etc. all vanish if $r = [1/2] \cdot [p_A - p_B + 1]$.

In the experiments reported $[1/2] \cdot [p_A - p_B] = 0.78$, which gives $C_1 = 0.00$ and $C_2 = 1.00$. If r is less than $[1/2] \cdot [p_A - p_B + 1]$, then to satisfy the equations one or more of C_3, C_4, C_5 etc. must be negative, which is physically impossible. If negative values are treated as zero and the value obtained from blood (0.72) is assigned to r, then $C_1 = 0.40$ and $C_2 = 0.60$. As the authors accept, these calculations must be interpreted with caution, but they do quite strongly suggest that nearly all the tumours were either monoclonal or biclonal.

This analysis depends on the fact that r differs significantly from 0.5. This is usual in mice and is a consequence of linkage between the PGK-1 locus and another locus on the X-chromosome of the mouse which is concerned with X-inactivation. No comparable situation has as yet been reported in other species so at present investigations of this kind will continue to be restricted to *mus musculus*.

3.3.2 Spatial distribution of clones

To determine the spatial distribution of different clones we would need markers recognizable at the cellular level. In principle, these might be surface molecules, like Ig or TATA, recognizable with monoclonal or conventional antibodies, or intracellular X-linked enzymes recognizable with antibodies or histochemical reagents.

In practice, little progress has been made towards developing appropriate antibodies. Histochemical reagents cannot be expected to distinguish between cells that produce different alloenzymes, but may distinguish between cells that produce normal amounts of a particular enzyme and those that produce little or none, and some work has been done on clonal distribution in hepatomas induced in mice heterozygous for deficiency of ornithine carbamoyl-transferase (OCT) (Howell et al. 1985).

3.3.3 Does the clonal composition remain constant?

As mentioned previously (Section 3.2.2, p. 34), lymphoblastoid cell lines from human B-cell lymphomas have been observed to differ from the cells of the original tumour in respect of two independent markers of N-clonality. It has been suggested by Fialkow (1972) that the cells of the new phenotype may have been derived from genetically foreign cells that gained access to the patient by transplacental passage from the mother, by blood transfusion, or via an insect vector. This sounds rather far-fetched, however, and it seems more plausible to postulate that the cells were derived from a neoplastic clone which was present in the original tumour but escaped detection because the cells were too few in number.

This would certainly seem to be the explanation of the changes in alloenzyme phenotype which were seen to occur (Woodruff et al. 1982) in fibrosarcomas that had been induced in heterozygous $Pgk-1^a/Pgk-1^b$ mice when they were grown in tissue culture or transplanted to histocompatible hosts, because the neoplastic nature of the cells exhibiting the new marker was confirmed by transplanting them to a homozygous female or hemizygous male of the opposite phenotype.

3.4 Tumour clonality in relation to carcinogenesis

Knowledge of tumour clonality is important for our understanding of carcinogenesis. Progress in this field has been limited partly by unsolved technical problems, but also by the failure of many writers on the subject to define clearly what kind of cell they have chosen as the founder cell of a neoplastic clone, or to adhere consistently to their definition (Woodruff 1988).

In the following discussion we shall use the terms *N-clonal*, *P-clonal* and *T-clonal* as defined in Section 3.1. As in Section 2.2, we shall, following Peto (1977), use *carcinogen* or *oncogen* to denote any external agent that increases the rate of entry into any given stage in the process of transformation or the probability that a transformed cell will proliferate successfully; and *oncogenic package* to denote the whole system of genetic and environmental factors involved in the development of a tumour, and their complex interactions. The genealogical tree that summarizes the cellular ancestry of a tumour will be termed the *pattern of carcinogenesis* (Woodruff 1988).

The factors that may be expected to influence tumour clonality are of three kinds:

1. The number and spatial distribution of susceptible target cells. This is important because, for the progeny of two or more cells to form one tumour, the founder cells must not be too far apart.
2. Various parameters relating to the oncogenic package and the growth potential of partly and fully transformed cells.
3. Interactions between different clones of partly and fully transformed cells, and between such cells and the host.

If there were no such interactions, and two identical individuals were exposed to the same oncogenic package, the probability that each would develop a single monoclonal tumour should equal the probability that one of them would develop either a biclonal tumour or two monoclonal tumours, depending on the volume of tissue in which the oncogenic package acts and the distribution of susceptible target cells. If, however, the development of one clone inhibits the development of another in the vicinity, the probability of a biclonal tumour will be correspondingly reduced, and if the inhibitory effect extends to clones anywhere in the same individual the probability of two separate monoclonal tumours will also be reduced.

A tumour would be likely to be N-monoclonal if the probability of transformation was low, or susceptible target cells were sparsely distributed, or if all clones except one were somehow eliminated. We cannot, without other evidence, distinguish between these possibilities, nor can we say anything about the pattern of carcinogenesis beyond the fact that the tumour is N-monoclonal. Two possible, but very different, patterns are illustrated in Fig. 3. In (a) the tumour is both N-monoclonal and T-monoclonal, whereas in (b) it is N-monoclonal but T-pleoclonal. The second possibility is something that many people do not seem to have envisaged but, in the writer's view, it is probably what happens when an N-monoclonal tumour develops in a field of precancerous cells.

If a tumour is N-pleoclonal it seems likely that the oncogenic package was such that the probability of transformation was relatively high, and that in at least one region there was a reasonably high concentration of susceptible target cells. It is, however, worth considering the possibility that as an N-monoclonal tumour develops it may promote the development of additional neoplastic clones or tumours by releasing growth factors which stimulate normal cells, or DNA which transfects them.

In the special case of tumours that are N-biclonal, there are two possibilities to consider. If the conditions are such that the probability of a cell becoming transformed is fairly small, the development of monoclonal tumours, together with a few that are biclonal, can be explained on stochastic grounds without

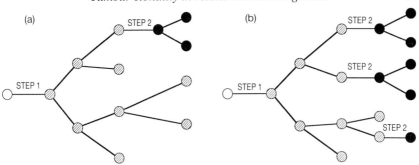

Fig. 3. Diagrams illustrating two different patterns of carcinogenesis. In both cases the tumour is assumed to develop in two steps from a single normal cell (open circle). In (a) the tumour arises from one transformed cell; in (b) from many. Hatched circles represent partly transformed cells; closed circles, fully transformed cells. (From Woodruff, M. F. A. (1988). *Advances in Cancer Research*, **50**. By permission of Academic Press.)

having to postulate clonal interaction. If, on the other hand, the probability that cells will become transformed and proliferate is high, a proportion of tumours might be expected to possess three or even more clones unless some were eliminated. Woodruff *et al.* (1986a) have suggested that, when a second clone of transformed cells arises in the presence of an established clone, a state of positive symbiosis may, or may not, develop. If it does, a third clone is unlikely to compete effectively against the established symbiotic pair; if it does not, one of the two clones is likely to disappear, after which the cycle may be repeated.

Progress in delineating patterns of carcinogenesis will depend very much on the extent to which we can develop new markers to distinguish cells at different stages on the way to transformation and sensitive methods for recognizing N-clonal, P-clonal, and T-clonal subpopulations at the cellular level, but even with the markers now available there is quite a lot that could be done.

The study of tumour clonality, like other areas of cancer research, still has much to gain from the use of appropriate animal models. As we have seen, sarcomas induced with methylcholanthrene have proved useful because the proportion of N-monoclonal tumours can be altered simply by altering the dose of carcinogen, and similar experiments could be set up with a variety of tumours induced with other agents. Students of B-cell lymphomas in transgenic mice bearing a c-*myc* oncogene coupled to the lymphoid-specific Ig heavy chain enhancer have extended the study of clonality to pre-neoplastic cells (Section 3.2.2, p. 35), and in so doing have opened up a new field that awaits further analysis.

With human tumours it would be of interest to compare the numbers of N-clones, P-clones, and T-clones at different stages of carcinogenesis. A start could be made with retinoblastoma and Wilms' tumour, and as new markers become available the study could be extended to tumours of other kinds. This might prove particularly rewarding with tumours in which there is a clinically recognizable pre-neoplastic or pre-invasive stage, such as carcinoma of the colon in patients with familial *polyposis coli*, primary carcinoma of the liver in patients with hepatic cirrhosis, and carcinoma of the cervix.

4
Dynamic heterogeneity of tumour cell populations

4.1 Ecological principles

The evolution of any population, whether of animals, plants or cells, depends on two essential elements:

1. Mechanisms for generating diversity among the members of the population.

2. Selection.

Ecologists have long been concerned with how animals and plants interact with their environment, and with each other; indeed, this is their *raison d'être*. Only comparatively recently, however, have tumour biologists become interested in studying tumour cell populations from an ecological point of view.

In this chapter we shall consider firstly how phenotypic differences between individual neoplastic cells in a tumour may arise, and secondly how the cellular composition of a tumour may be affected by the reaction of the host, and by interactions between subpopulations of tumour cells. The effect of treatment is, of course, also relevant, but it will be convenient to defer consideration of this topic until Chapter 6.

4.2 The genetic basis of phenotypic diversity

The factors which may give rise to heritable phenotypic differences between the cells of an individual may be classified as follows:

1. Mutations affecting genes that directly determine a particular phenotypic character.

2. Changes in gene expression resulting from:
   (a) Mutations in other genes.
   (b) Insertional mutagenesis.
   (c) Factors outside the genome, in the cell nucleus, the cytoplasm or the extracellular environment.

The term *mutation* is used in a broad sense to denote any heritable change in the genetic material, and includes point mutations, deletions, insertions, rearrangements and gene multiplication, *Insertional mutagenesis* means that a gene is activated as the result of insertion of new genetic material in its vicinity, either by integration of a retrovirus or as the result of a chromosomal translocation. It is convenient to have a collective term to denote the operation of all factors outside the genome, and, despite its chequered history, we shall use *epigenetic* for this purpose.

Before discussing how these various factors may operate to generate diversity among the neoplastic cells of a tumour, it may be helpful to look first at how they operate in normal development. Even 20 years ago little could be said about this, and the question of how a highly organized assembly of specialized cells and tissues develops from a single fertilized egg cell looked as if it might prove insoluble, at any rate for a very long time. Today, thanks in large measure to the growth of molecular genetics, the situation is very different, and those engaged in the study of cancer can learn much from developmental biologists.

4.2.1 Normal development in eukaryotes

Mutation is important in the development of the immune system, where recombination between different germ-line sequences and somatic mutation both play an important role in generating the great cellular diversity required for the recognition of a vast range of antigens by T and B lymphocytes, and the production of specific antibodies by B lymphocytes and plasma cells derived from them (see Baltimore 1981; Tonegawa 1983; rev. Roitt *et al.* 1985).

Apart from this, the emergence of specialized cells and tissues during development appears to depend almost entirely on epigenetic mechanisms. Even terminally differentiated cells commonly retain the full diploid complement of genes, and some remain 'totipotent', at least for a considerable time, in the sense that inactive gene sequences can be reactivated, or can give rise to activated sequences following DNA replication and mitosis (see Maclean and Hall 1987). The first evidence of this was provided by the demonstration that the cell formed by transplanting the nucleus of a tadpole intestinal epithelial cell to the cytoplasm of an enucleated frog's egg could develop into a normal adult frog, and that a cell formed by transplanting the nucleus of a somatic cell from an adult frog to an enucleated egg cell could develop as far as the tadpole stage (see Gurdon 1977). More recently, the totipotency of mammalian embryo cells has similarly been demonstrated by nuclear transplantation (Illmensee and Hoppe 1981).

All genes, including 'housekeeping genes' that are active in a wide variety of cells, are subject to some degree of regulation, but in discussing development we will be concerned primarily with genes coding for tissue-specific cell proteins. In cells where a particular product would be inappropriate the corresponding genes are not transcribed, largely as the result of condensation of chromatin, which makes it inaccessible to RNA polymerase, and methylation of cytosine residues. Changes in the level of transcription may be brought about by gene regulating molecules produced by regulatory genes. Gene expression may also be regulated at the transcriptional, translational or post-translational level by signals from the cytoplasm or the cell surface. These signals may be triggered by factors in the extracellular environment, including hormones, extracellular diffusible morphogens, contact with neighbouring cells and the extracellular matrix, and substances entering the cell from neighbouring cells through gap junctions (rev. Fristrom and Spieth 1985; Maclean and Hall 1987).

The hypothesis that cells continuously produce a variety of gene regulating molecules that allow the expression of only those genes appropriate for the type of cell in question, gains support from studies of hybrids between highly specialized cells like hepatocytes, melanocytes and immunoglobulin-producing cells, and less specialized fibroblasts, in which the more differentiated phenotype is usually extinguished but may reappear after chromosome loss (see Bernhard 1976; Klein and Klein 1986).

So far, all gene regulatory molecules appear to be proteins but, as Maclean and Hall have pointed out, there seems no reason why RNA should not function in this way.

As a rule, changes in gene expression do not occur in the absence of DNA synthesis, but an exception, in an experimental situation, has been reported by Chiu and Blau (1984), who observed that a human form of the contractile enzyme, creatinine kinase, was produced when human fibroblasts were fused *in vitro* with mouse muscle cells in the presence of an inhibitor of DNA synthesis.

It has long been known that interaction between cells of different lineage plays an important role in development, and is responsible for what used to be called *embryonic induction*. Interaction between mesenchymal and epithelial cells also occurs in adults, notably in the small bowel, where epithelial cells and sub-epithelial mesenchymal cells migrate together from crypts to villi, and in the colon, where joint migration of epithelial and mesenchymal cells also occurs although there are no villi (rev. Wessels 1977).

More recently, it has become apparent that interaction also occurs between cells and the extracellular matrix (ECM), or ECM components such as collagen, glycoproteins and glycosamino glycans (GAG), and enables the ECM to contribute to the regulation of gene expression.

Bissell and her colleagues (Bissell *et al.* 1982; Bissell 1988) have investigated the matter by comparing the morphology of mouse mammary epithelium, and its capacity to synthesize and secrete milk proteins, when cultured (a) on plastic in normal medium, (b) on fixed or floating collagen gel, and (c) on plastic in medium containing various ECM components. Cells cultured on plastic in normal medium became abnormal in shape, and their capacity to secrete milk proteins was very much reduced, though they responded to some extent to lactogenic hormone. They remained distorted on fixed collagen gel, but under the other conditions listed their shape was much more like that which occurs *in vivo*, and the secretion of proteins such as transferrin and beta-casein, and the response to lactogenic hormone, were greatly increased. It was concluded that there is a complex interplay between hormones, the ECM and cell 'shape' which determines whether or not a given milk protein is synthesized and whether it is secreted. Further experiments using human and mouse c-DNA clones and monoclonal antibodies to various milk proteins point to the conclusion that the ECM influences gene expression as a consequence of transmembrane connections with the cytoskeleton, and the cell in turn influences the structure and composition of the ECM. Bissell *et al.* suggest that this 'dynamic reciprocity' between a cell and the ECM is necessary for the stability of the differentiated state.

Normal differentiated cells, as we have seen (Section 2.1), can undergo only a limited number of cell divisions *in vitro*. The permitted number shows a positive correlation with the mean life span of the species, and an inverse correlation with the age of the individual, from which the cells were obtained. It is reduced in patients with Werner's syndrome, who undergo premature ageing (Halliday *et al.* 1974). These findings suggest that the number of doublings of differentiated cells may also be limited *in vivo*.

The situation is quite different with the germ line cells that are set aside to provide the future gametes, and with stem cells in various tissues that can give rise to similar stem cells, or differentiate into one, two, or occasionally more, kinds of specialized cell.

4.2.2 Generation of cellular diversity in tumours

Sometimes, as we have seen (Chapter 3), a tumour develops from more than one normal cell. Possible ways in which this might contribute to the diversity of the neoplastic cells in the tumour have been examined in Section 3.2.1, pp. 32–33).

Most tumours, however, are N-monoclonal and, like the developing embryo, pose the problem of how a diversified population develops from a single progenitor cell. While the progenitor cell of an embryo and a tumour may differ little in their genomic content, however, they differ markedly in the extent to which genes are expressed, and the relative importance of the various mechanisms which determine the parameters of the emergent population of cells is also very different.

Carcinogenesis appears to depend on two processes:

1. Mutation (in the broad sense of the term) in a normal gene, which makes it function as a cancer gene.

2. Activation or enhanced activity of an otherwise harmless normal gene. There are, as we have seen (Section 2.4.1, pp. 13–14), many well-known instances

in which this occurs as the result of chromosomal translocation or the integration of a retrovirus.

Mutation is also involved in the natural progression of tumours, and in the adaptation of tumour cells to environmental changes, including those resulting from treatment. Neoplastic cells appear to be more susceptible than normal cells to chromosome breakage, failure of DNA repair, non-dysjunction and polyploidy, sister chromatid exchange, gene multiplication, and other genetic aberrations, both *in vitro* and *in vivo* (Nowell 1976, 1982). This genetic instability, as Nowell has pointed out, may be due to inherited or mutationally acquired defects in DNA repair mechanisms, or continued exposure to environmental mutagens.

Some oncogenes, especially c-*myc* and N-*myc* but also members of the *ras* family, c-*erb B* and other oncogenes, may be multiplied dozens, or even hundreds, of times in certain tumours, and this is regularly associated with a high level of transcription (rev. Alitalo and Schwab 1986). Gene multiplication of this order, sometimes accompanied by translocations and DNA rearrangements, is typically associated with increasingly aggressive behaviour of the tumour characteristic of natural progression (see Klein and Klein 1986). This is illustrated, so far as human tumours are concerned, by the following examples:

(1) In small cell lung cancer (SCLC), as mentioned earlier (Section 2.4.1, pp. 17–18), high amplification of c-*myc* has been found in a cell line classed by Minna and his colleagues as 'variant' on account of altered morphological and biochemical features in tissue culture, increased tumorigenicity in athymic nude mice, and the fact that the tumours from which the lines were derived behaved clinically in an even more malignant way than typical SCLC (Little *et al.* 1983; Nau *et al.* 1984).

(2) Amplification of c-*myc* has been found in plasma cell leukaemia, which is the most malignant form of human plasma cell tumour, but was absent in a large number of solid plasmacytomas (Sumegi *et al.* 1985).

(3) Amplification of N-*myc* occurs in tumour tissue and cell lines from advanced neuroblastomas and retinoblastomas but has not been found in less advanced tumours. Neuroblastoma cells can be induced to differentiate morphologically and biochemically *in vitro* by treatment with various agents including retinoic acid. Thiele *et al.* (1985) found that expression of N-*myc* was decreased 6 hours after addition of retinoic acid, before changes in morphology or in the cell cycle were apparent. After 48 hours, when N-*myc* expression was maximally depressed, cell cycle arrest was also maximal, 14 per cent of the cells treated with retinoic acid being in the growth fraction as compared with 37 per cent of those in control cultures.

Gene multiplication may also occur in a wide range of tumours, and occasionally in normal cells, when there is clear evidence of selective pressure occasioned by environmental changes; in particular, exposure to cytotoxic agents. It has been shown that various segments of DNA may be replicated more than once during a single cell cycle, and it has been plausibly suggested that this illegitimate DNA is normally lost but may be retained under conditions of selective pressure. The incidence of amplified genes may be greatly increased if cells have been exposed previously to a carcinogen, or if they are exposed to mitogens while the selective pressure is operating (rev. Alitalo and Schwab 1986).

Gene amplification in cultured tumour cells is typically associated with large, cytologically visible regions that stain homogeneously in the relevant chromosomes, and the presence of numerous double minute fragments. These features may be less apparent in tumour cells that have not been cultured *in vitro*, and in all cases the precise situation can only be established by means of appropriate molecular probes.

Activation of a gene by epigenetic factors of the kind that are so important in normal development does not appear to play a role of comparable magnitude in carcinogenesis, though it seems possible, as Bissell (1988) has suggested, that disruption of the interaction between a cell and the extracellular matrix may lead to destabilization of the differentiated state and thus contribute to the development of some carcinomas.

Epigenetic factors may conceivably play a bigger role in tumour progression and the generation of cellular diversity (rev. Schirrmacher 1985). This suggestion gains support from the fact that some neoplastic cells secrete growth factors, and that transformed cells have been shown, though as yet only *in vitro*, to communicate with other transformed cells via gap junctions (Sections 1.2.2 and 2.6.1, p. 26).

As we have seen, stem cells, which can give rise either to more stem cells or to a limited range of differentiated cells, play an essential role in maintaining differentiated tissues in which there is a high rate of cell turnover. In the case of a tumour with a heterogeneous, but fairly stable, population of neoplastic cells, three possibilities would seem to merit consideration:

1. There is a single category of stem cell by which the whole population is maintained.

2. Some, at least, of the subpopulations possess their own distinctive stem cells.

3. Apart from a relatively small number of doomed cells (Section 1.2.2), most, if not all, of the neoplastic cells in the tumour behave as stem cells.

In theory, at least, it should be feasible (see Section 1.3.1) to distinguish between these possibilities by collecting samples of viable tumour cells that differ in respect of DNA content or some other convenient parameter by flow cytometry, and looking to see what kinds of cells (if any) they give rise to *in vitro* and *in vivo*.

4.3 Cell cycle dynamics

The neoplastic cells of a tumour, as has been pointed out already (Section 1.2.2), include proliferating cells in various stages of the cell cycle, G_O cells that have stopped cycling temporarily but may later re-enter the cycle, doomed cells that are incapable of further division, and dead cells.

Methods have been developed for estimating, *inter alia*, (1) the proportion (f_G) of proliferating cells; (2) the duration (T_C) of the cell cycle for the proliferating cells, and its component phases; (3) the tumour cell birth rate (k_B); and (4) the rate of cell loss (k_L). These methods, the errors to which they are subject, and the ethical considerations that limit their use in humans, have been the subject of numerous reviews (see, e.g. Ahern *et al.* 1980). In brief, they include (a) determining the proportion of labelled cells, and labelled mitoses, on autoradiography after an injection of ^{3}H-thymidine; (b) determining the proportion of cells in metaphase before and after injection of a stathmokinetic agent (i.e. an agent which arrests the cycle when a cell reaches metaphase); and (c) monitoring the disappearance of label from a tumour after injection of ^{3}H-thymidine, ^{125}I- or ^{131}I-deoxyuridine, which, if ^{131}I is used as the label, can be done by external counting.

These methods provide information about the neoplastic cell population as a whole; they do not distinguish between subpopulations of different lineage for which, at any given time, the actual, though unknown, values of f_G, T_C, k_B and k_L may well be very different. Moreover, the ways in which these parameters change with time are likely to differ for different subpopulations, because these may respond differently to changes in the host environment, including those caused by treatment, and interact in complex ways with one another.

4.4 Dynamics of lineage heterogeneity

4.4.1 Influence of the host

The relationship between a tumour and the host in which it originates may be described as one of *symbiosis*, using this term in its original sense of *living together*, irrespective of the advantages and disadvantages of the relationship to either partner. Although some tumours that are classified as malignant on account of their histological structure, or their behaviour in tissue culture, appear to have little or no harmful effect on the host for months or years, this is exceptional; much more commonly the malign effects of what we call malignant tumours are only too apparent. The effect of the host on the tumour, however, is often much less clearly perceived, and this gave rise to the dogma, which was for a long time accepted uncritically, that malignant tumours are autonomous. The meaning that different people attach to *autonomous* is not always clear, but in general the term seems to imply, firstly, that tumours are not influenced by any of the factors that control the growth and equilibrium of normal tissues, and, secondly, that no special homeostatic mechanisms have evolved for controlling carcinogenesis and tumour growth.

The first proposition became untenable when Charles Huggins showed unequivocally that the growth of some tumours in animals and humans is influenced by hormones. As he said when reviewing this work in his Nobel Lecture (Huggins 1967), 'cancer is not necessarily autonomous and intrinsically self-perpetuating. Its growth can be sustained and propagated by hormonal function in the host which is not unusual in kind or exaggerated in rate but which is operating at normal or even subnormal levels.'

Tumours may be induced by hormonal imbalance; when this happens they typically remain for a time dependent on the continued existence of the state of imbalance but, sooner or later, a population emerges that can proliferate in a hormonal normal environment, and this change is likely to occur more quickly if the degree of imbalance declines. Jacob Furth, who inaugurated the study of this process (see Klein and Klein 1986), described the change as progression from a *conditioned* to an *autonomous* state, and predicted that a comparable process might occur with neoplastic cells regulated by homeostatic mechanisms of a different, albeit unknown, kind. The existence of such mechanisms, though still occasionally disputed, seems now beyond doubt (see Woodruff 1982), and confirmation of Furth's prediction has been provided by the observation that tumours which are susceptible to attack by cytotoxic T-cells or NK-cells may become progressively less sensitive, as shown by their changed behaviour in the autochthonous host and the increasing ease with which they can be transplanted to normal syngeneic hosts (see, for example, Woodruff 1986*b*; Woodruff *et al.* 1986*b*).

Klein and Klein (1986) have taken Furth's suggestion still further, and have re-examined the concept of conditioned and autonomous neoplasms in the light of current knowledge of oncogene activation. In particular, they have postulated that *myc* activation renders cells more autonomous by making them less sensitive to signals that ordinarily induce differentiation or limit growth. This hypothesis is consistent with the association of c-*myc* and N-*myc* amplification with clinical tumour progression, examples of which were discussed in Section 4.1.2, and with the observation of Thiele *et al.* (1985), that the growth of neuroblastoma cells in tissue culture is inhibited by treatment with retinoic acid, which in this system inhibits transcription of N-*myc*. It is noteworthy that the translocated *myc* oncogene in Burkitt's lymphoma cannot be turned off in this way; it seems likely, therefore, that retinoic acid would not inhibit the growth of Burkitt's lymphoma cells, but this prediction does not seem to have been tested experimentally.

From what has been said, it is easy to see why, and in some cases how, neoplastic cells adapt to unfavourable and changing conditions in the host environment. But why do the neoplastic cells of a successful tumour remain so heterogeneous? The answer, as Heppner (1984) has argued, lies in the nature of the evolutionary process, which is based on selection between interacting populations rather than between alleles at each locus separately. As a rule, each phenotypic character is affected by many gene substitutions, and each substitution affects more than one phenotypic character; what is optimal for a population may therefore be given by several different genetic combinations, and the observed phenotypes display what Sewall Wright (1982) has called 'a shifting balance'.

Even today, the fact that tumour cell subpopulations interact is sometimes overlooked, and people argue that, given two subpopulations that grow independently and exponentially at different rates, one will always be swamped by the other and disappear. The reasoning is sound; the conclusion is wrong because the premise is false.

4.4.2 Interactions between tumour cell subpopulations

The experimental study of interactions between subpopulations of tumour cells has been made possible by the existence of markers that permit recognition of cells belonging to particular subpopulations, and the development of cloning techniques that enable cell lines to be established from single cells.

In this section we shall discuss the following topics:

1. The dynamic heterogeneity of cloned and uncloned tumour cell populations *in vitro* and *in vivo*.

2. Experiments with combinations of cloned tumour cell populations.

3. The limitations of studies with cloned tumour cell subpopulations.

Dynamic heterogeneity of cloned and uncloned populations

It has been reported from various laboratories that subpopulations isolated from tumours by *in vitro* cloning usually become heterogeneous if they are maintained in tissue culture or transplanted to a suitable host (rev. Heppner 1982, 1984; Fidler and Hart 1982).

The property that has been most studied in this context is the capacity to produce what are often, though in the writer's view unfortunately, called *artificial metastases*: i.e. disseminated tumour foci that develop after intravenous injection of tumour cells. It has been shown with the B16 mouse melanoma (Poste *et al.* 1981; Hill *et al.* 1984), and with a mouse fibrosarcoma (Harris *et al.* 1982), that, when clones that are 'non-metastatic' in this sense are grown in tissue culture or transplanted, metastatic variants soon appear, and conversely. Moreover, in the case of the B16 melanoma, the rate at which 'metastatic' variants were generated in culture differed for two sublines of the same tumour.

In experiments with a murine lymphoma, Chow and Greenberg (1980) found that heterogeneity in respect of susceptibility to killing by naturally occurring cytotoxic antibody developed during a single passage *in vivo* but did not develop *in vitro*; and concluded that host factors had somehow contributed to the generation of diversity.

Some cloned lines are more stable than others. This is seen in experiments by Heppner's group (rev. Heppner 1982) with five cloned sublines of a mouse mammary carcinoma which differed in karyotype, growth characteristics *in vitro*, tumorigenicity on subcutaneous injection, 'metastatic' capacity on intravenous injection, immunological features, and sensitivity to cytotoxic drugs and irradiation. These sublines were used mainly in experiments with combinations of subpopulations, as described in the next section, but it was noted that one subline was particularly unstable *in vitro* and gave rise to a whole series of sub-subpopulations differing in karyotype, morphology and growth characteristics, whereas the other four were much more stable.

As a general rule, in contrast to the instability of cloned subpopulations, the composition of the original uncloned populations remains much more constant under similar conditions of culture or transplantation. Poste *et al.* (1981), who first reported this, suggested that in heterogeneous populations the various subpopulations interact in such a way as to stabilize their relative proportions within the population as a whole. The composition of uncloned populations may, nevertheless, change quite suddenly, presumably in response to an environmental change, though this may not be identified. This phenomenon is illustrated by the experiments of Woodruff *et al.* (1982) with fibrosarcomas raised in mice heterozygous for the alloenzymes (A and B) of PGK-1, which were described in Section 3.3.3. The relative proportions of the two alloenzymes in cell digests changed markedly in the course of long-term tissue culture, and sometimes one form which could not be detected initially became predominant, or even the only form detected. Moreover, in one experiment, one alloenzyme became undetectable after repeated transplantation but subsequently reappeared. The probability of these results being due to random mutations is minuscule, and the conclusion seems inescapable that sparse populations may persist for a long time both *in vitro* and *in vivo*, and suddenly begin to expand rapidly. It seems likely that most, and possibly all, of the cells in these sparse populations are in G_O, but the possibility that they are cycling slowly, and that cell birth is balanced by cell death, is not excluded.

Experiments with combinations of cloned subpopulations

Heppner and her colleagues (rev. Heppner 1982) studied the behaviour of cultures set up with mixtures of two cloned subpopulations of a mouse mammary carcinoma, one of which expressed the mouse mammary tumour virus (MuMTV) antigen. The proportions of the two components remained constant, though they grew at very different rates when cultured separately; and the growth rate of the mixture was that of the faster component, or that of the slower, depending on the culture conditions. The composition of clonal mixtures may, however, change on repeated subculture; in the writer's laboratory, for example, in a culture established with a mixture of equal numbers of cells of two clones of a murine fibrosarcoma, one of which expressed PGK-1 A and the other PGK-1 B, which was subcultured every seven days, the A/B ratio fluctuated around a mean value of 3/1 for seven weeks, but thereafter the B component became undetectable.

Heppner *et al.* showed also that when two subpopulations were plated on separate coverslips, and these were immersed in medium in the same dish, some populations inhibited the growth of others during the period of observation (three days); in contrast to the findings with mixed cell populations, however, no stimulatory effects were observed. When two populations which differed in sensitivity to methotrexate were grown in the same way on separate coverslips, in the presence of methotrexate in appropriate concentration, the sensitivity of the less sensitive population sometimes increased, though this did not occur with all combinations of sensitive and insensitive populations studied.

In other experiments, Heppner's group injected mice subcutaneously at one site with a subpopulation of a mouse mammary carcinoma that was sensitive to cyclophosphamide (Cy), and at another site with a subpopulation from the same tumour that was Cy-insensitive. Different groups of mice received different doses of Cy. At a dose level at which the subpopulations showed a clear difference in sensitivity when tested separately, the insensitive population became increasingly sensitive in the mice inoculated with both subpopulations. This still occurred if the mice were irradiated prior to the injection of tumour cells.

Limitations of studies with cloned subpopulations

It is important to realize that the behaviour of subpopulations isolated by cloning, or a mixture of such subpopulations, may not accurately reflect the behaviour of cells of the same lineage in the original tumour.

In the writer's laboratory it has been observed repeatedly that many *in vitro*-propagated cloned cell lines, and some uncloned lines, from strongly immunogenic murine fibrosarcomas, fail to grow when transplanted subcutaneously to normal histocompatible hosts, whereas cells from tumours that have not been propagated *in vitro* are readily transplantable. Reduced tumorigenicity of various other tumours propagated *in vitro* has also been reported (rev. Boon 1983), associated sometimes, but not always, with increased immunogenicity.

In the case of the fibrosarcomas, cell lines that failed to grow in normal adult histocompatible mice grew readily in various categories of T-cell deficient mice, including syngeneic or allogeneic congenitally athymic (nude) mice, and after such passage grew readily in normal histocompatible mice. The non-transplantability of the cultured cell lines suggests that a phenotypic character that is essential for growth *in vivo*, but irrelevant or possibly even disadvantageous *in vitro*, is lost during tissue culture, especially if the first step in establishing the line was cloning by limiting dilution.

Four hypotheses were proposed to account for the recovery of tumorigenicity during passage *in vivo* in a T-cell deficient host and tested experimentally, namely: (1) The mouse-passaged cells did not express tumour associated or MHC antigens; (2) Some at least of the passaged tumour cells were resistant to NK cells (Woodruff 1986*b*), and cellular immunity mediated by sensitized T-cells in itself was not sufficient to destroy them; (3) During passage in the immunodeficient mouse the cells acquired a protective surface molecule that interfered with the efferent side of the immune response when the cells were retransplanted to a normal mouse; (4) The mouse passaged cells, but not cells taken directly from tissue culture, produce sufficient amounts of growth factors to permit growth *in vivo* (Woodruff and Hodson 1985*a*, 1985*b*; Woodruff *et al.* 1986*b*).

The first hypothesis has been definitely excluded, but the evidence suggests that the other three, either individually or in combination, may play a role. Whatever the explanation, the phenomenon highlights the need for caution in interpreting observations made with cloned cell lines. For some purposes it may suffice to passage the line *in vivo* in a suitable host before it is used; for studying clonal interaction, however, this would be self-defeating because, even if the line was of sufficiently recent origin to still be homogeneous, it would quickly become heterogeneous on passage *in vivo*.

4.5 Tumour progression and regression

Analysis of tumour cell populations shows clearly that they are heterogeneous, and that this heterogeneity is not static but dynamic. Experimental observations show that cloned subpopulations diversify, and interact with other subpopulations both *in vitro* and *in vivo*. While due allowance must be made for the effect of cloning and culture *in vitro*, it seems inconceivable that interactions should not also occur in whole tumours, where intercellular competition may be intense and each cell can contribute to the common environment in various ways, including the secretion of factors that stimulate (Todaro *et al.* 1982) or inhibit growth. Indeed, clonal interaction provides the only convincing explanation of the finding, discussed in Section 3.4, that even under conditions of carcinogenesis where the probability of a tumour developing is high, the tumours which arise are often N-monoclonal, and probably rarely, if ever, more than N-biclonal.

Interaction may result in the complete elimination of one or more subpopulations, but, as we have seen, it has been shown that very sparse subpopulations may persist for a long time, both *in vitro* and in transplanted tumours, and there seems no reason to doubt that this may occur also in autochthonous tumours.

Dynamic heterogeneity is associated clinically with tumour progression. Progression may be arrested at any stage, and sometimes, in both animal and human tumours, there is a change in the opposite direction, as shown by partial or complete spontaneous regression of a tumour (rev. Woodruff 1980). In this context, the term *spontaneous* means that the regression could not be attributed to treatment, either because no treatment was given, or because the patient received only palliative treatment of a kind which does not normally result in tumour regression. In some cases non-specific immunostimulation resulting from infection, or changes in hormonal balance, may have played a role; in others, however, it is impossible to make even a plausible guess as to why regression occurred.

Opportunities for studying spontaneous regression of human cancer are limited because the natural history of the disease is, fortunately, often profoundly modified by treatment, but some patients present too late for anything other than palliation, and others refuse treatment. Because of this, Everson and Cole (1966) were able to identify 176 cases reported in the literature between 1900 and 1966 which, in the light of quite strict criteria, they regarded as possible examples of partial or complete regression, and further examples were discussed more recently at a conference at the US National Cancer Institute (National Cancer Institute 1976).

In 64 of the patients reviewed by Everson and Cole there was regression of a primary tumour following biopsy or manifestly incomplete excision. Some of these patients were alive and appeared clinically to be free of tumour; others had died and no tumour was found at autopsy. Others again showed no tumour, a small amount of recognizably malignant tumour, or apparently benign residual tumour (Section 2.6.2), at a subsequent operation. Regression is more common with some kinds of tumour, especially neuroblastoma, malignant melanoma and chorion carcinoma, than with others, but there are some well-documented instances of regression of carcinomas of the colon, stomach, liver, pancreas, bladder, lung, larynx, thyroid and other sites, and of soft tissue sarcomas. In 5 of 19 patients with neuroblastoma in the series reviewed by Everson and Cole, regression took the form of a change in histological appearance to that of benign ganglioneuroma, and further instances of this change have been reported more recently (see Thiele *et al.* 1985).

Further evidence of interaction between tumour cell subpopulations, of the persistence and sudden expansion of sparse populations, and of tumour regression, is provided by the study of metastasis, which forms the subject of the next chapter.

5
Invasion and metastasis

Many cells exhibit migratory activity, i.e. can crawl *in vitro* and *in vivo*, because they possess contractile fibrils and have the capacity to form surface adhesions of various kinds. This phenomenon has been studied in depth by Abercrombie and his colleagues (rev. Abercrombie 1980). Crawling, according to Abercrombie, subserves three functions: (1) in the case of wandering cells (leucocytes and macrophages), defence against micro-organisms; (2) maintenance of organization during growth, and restoration of organization after tissue damage; and (3) translation of embryonic determination during development into the definitive arrangement of cells characteristic of the mature organism. Crawling activity is stimulated when there is rapid cell division, and is influenced also by the cell's environment. In the case of normal cells, crawling is arrested when the leading edge of one cell comes in contact with some part of another cell—a phenomenon termed *contact inhibition* (Abercrombie and Heaysman 1954)—but may be resumed later in a different direction. It is also arrested by epithelial basement membrane, and directed by extracellular connective tissue matrix.

The growth of benign tumours resembles that of normal tissues. Such tumours displace, but do not invade, surrounding tissues. Malignant tumours are relatively independent of contact inhibition and other factors that control the migration of normal cells. While they may displace neighbouring structures, they spread locally mainly by *invasion*, and may also give rise to secondary foci of invasive growth at a distance from the main tumour. These are termed *metastases*, and the process is termed *metastasis*. It is generally accepted that metastases are derived from viable cells which have come either from the primary tumour or from another metastasis but, while this is certainly the rule (see Tarin 1985), it seems wise to keep an open mind as to whether some metastases are due to transfection of local cells by disseminated subcellular material (Woodruff 1986a).

In this chapter we will examine the biological features of invasion and metastasis, and, in particular, the importance of cellular variation and adaptation in this context. The clinical significance of these processes will be considered in Chapter 6.

5.1 Analysis of invasive growth

Invasive growth involves disruption of epithelial basement membranes and extracellular connective tissue matrix, and cellular infiltration of surrounding normal tissues. Some tissues appear to favour the local spread of particular tumours—a basal carcinoma of the skin, for example, typically spreads more rapidly after it has invaded bone; other tissues, like cartilage, resist invasion by most tumours.

Three mechanisms have been postulated (see Schirrmacher 1985; Liotta *et al.* 1987) to account for invasive behaviour:

1. Mechanical pressure generated by the expanding tumour cell population.

2. Lysis by proteolytic enzymes. Tumour cells attach to the laminin of the basement membrane via cell-surface laminin receptors and then secrete enzymes that can degrade Type 4 collagen. Proteases secreted by stromal polymorphonuclear leucocytes may possibly contribute to the process.

3. Abnormal migratory activity of tumour cells. This depends on a complex set of properties associated with changes in the cytoskeleton of the cell which have been referred to collectively as the *locomotor phenotype*. They include the ability of the tumour cells 'to adhere to structures present on the invasion front' and to change shape (Zimmerman and Keller 1987).

These mechanisms are, of course, not mutually exclusive, and it seems likely that all three are often involved.

What concerns us particularly in this book is that dynamic heterogeneity is likely to be important for successful invasion. This is because the nature of the barriers encountered, and in consequence the phenotypic features needed to overcome them, may be different at different sites.

5.2 Routes of metastasis. Analysis of the process

As a background to recent work on metastasis, the reader is referred to Rupert Willis's (1952, 1973) classic monograph, *The spread of tumours in the human body*, and Warren Cole's (1973) review, *The mechanisms of spread of cancer*. Following Cole, we shall discuss the dissemination of cells leading to metastasis under the following headings:

1. Dissemination via lymphatics

2. Dissemination via the bloodstream

3. Implantation of tumour cells on serous or epithelial surfaces, and in wounds.

Tumour biologists sometimes use the term metastasis in the restricted sense of metastasis via the bloodstream, or extend it to include the development of scattered tumours after experimental intravenous injection of viable tumour cells. Both practices are to be deprecated because they present an oversimplified view of a complex process.

5.2.1 Metastasis by lymphatics

Malignant tumours often invade lymphatics, and tumour cells or cell clumps may then be carried to the regional lymph nodes. This is termed *lymphatic embolism*. Another process, termed *lymphatic permeation*, has been described (Handley 1922) in which tumour tissue grows as a column within the lumen of lymphatic vessels; this is less common than lymphatic embolism but does sometimes occur.

Some tumour cells that gain entry to lymphatics are arrested in the first node they reach, and may there give rise to a metastasis. The arrest and survival of tumour cells in lymph nodes has also been demonstrated experimentally in rats by injecting tumour cells into an afferent lymphatic of a large lymph node and then transplanting fragments of the node to a susceptible host. These experiments have shown also, however, that the 'lymph node barrier' is less efficient than was originally believed, and that many of the injected cells pass through the node to the efferent lymphatic, and thence, via the thoracic duct, to the bloodstream (Fisher and Fisher 1966a,b). Other experiments (Hewitt and Blake 1975), in which the nodes draining the site of a subcutaneous or intradermal transplant of a murine squamous cell carcinoma were transplanted to the same mouse, or to another mouse of the same strain, point to the same conclusion. The nodal transplants gave rise to tumours in 40 per cent of cases, but this was attributed to the presence of tumour cells in transit through the node because, when the nodes were left *in situ* after radical excision of the primary transplant, metastases developed in the nodes in only 4 per cent of mice.

5.2.2 Metastasis by the bloodstream

The development of a bloodborne metastasis involves a series of steps, sometimes referred to as a cascade:

1. Invasive growth of the primary tumour, or another metastasis, in the vicinity of a blood vessel.

2. Penetration of the vessel wall—typically a small vein—by tumour cells. It is often assumed that veins are more susceptible than arteries to invasion by tumours because their walls are thinner, but differences in intraluminal pressure may also be important (Shivas and Finlayson 1965).

3. Transport of tumour cells in the bloodstream, either as single cells or as clumps, and their arrest by embolization of a small vessel.

4. Penetration of the vessel wall.

5. Development of a vascularized stroma.

6. Invasive growth of tumour cells into the surrounding tissue.

Alternatively, instead of steps 1 and 2, tumour cells may gain access to the bloodstream via the lymphatic system, as described in Section 5.2.1.

Human tumours

Observations in living patients and at autopsy (rev. Willis 1952, 1973; Cole 1973; Woodruff 1980, pp. 52–8; De la Monte *et al.* 1983; Schirrmacher 1985) point to the conclusion that the incidence and distribution of metastases depend on the following factors:

1. Anatomical factors which influence the distribution of tumour emboli.

2. The number of tumour cells arrested, and the extent to which they are aggregated in clumps.

3. Properties of the new environment—referred to by Paget (1889) as the *soil* in which the cells are seeded—which determine its capacity to satisfy the metabolic and other requirements of the tumour cells.

4. Properties of the tumour, including its capacity to produce plasmin-like enzymes.

5. Various factors which influence local and general resistance to the tumour.

When the primary tumour is situated in a region whose venous drainage is via the portal vein, the site at which circulating tumour cells first encounter a capillary bed is the liver; when the tumour is situated elsewhere it is normally the lungs. Tumour cells may, however, undergo sufficient deformation to pass through capillaries, or may by-pass the capillary bed and enter the venous side of the circulation via arteriovenous shunts in the liver, lungs and elsewhere, and then enter the arterial circulation via the heart, and gain access to vascularized tissue anywhere in the body. Metastasis to bone sometimes occurs by retrograde venous flow when the venous pressure is raised, for example during coughing or sneezing. A familiar example is metastasis of carcinoma of the prostate to the bones of the pelvis and vertebral column, via the vertebral venous plexus of Batson.

In pregnant women metastases develop occasionally in the placenta, and very occasionally in the fetus.

It is noteworthy that metastases in the liver may derive their main blood supply from the hepatic artery, but this is not always the case and the pattern of vascularization may change as the tumour grows; similarly, metastases in the lung sometimes derive their blood supply from the bronchial artery.

The presence of cancer cells in the blood of patients with primary tumours has been reported by many observers. It might be expected that the demonstration of cancer cells in blood samples taken during an operation for removal of a primary tumour would be associated with an especially poor prognosis, but this is by no means always the case. The most remarkable observations of this kind are those reported by White and Griffiths (1976), who found an inverse correlation between the presence of tumour cells in venous blood samples collected during resection of colorectal carcinomas and the subsequent development of hepatic metastases. Great care was taken to avoid erroneously diagnosing normal cells as neoplastic, and more than 200 patients were followed for 10 years. It seems clear that many of the cells that gained entry to the bloodstream failed to form metastases.

Analysis of bloodborne metastasis in animals

Studies in animals of the distribution of tumour colonies after injection of tumour cells to a systemic vein or the portal vein, including in some experiments cells labelled with ^{131}I-deoxyuridine, point to the conclusion that this too depends partly on anatomical factors and partly on the suitability of the new environment. They have shown further that only a small proportion—typically 1 per cent or less—of the injected cells give rise to colonies in the organ of first encounter; and that, while many of the cells that pass through and re-enter the circulation are also destroyed, some survive and form colonies elsewhere (Weiss 1980*a*; Weiss *et al*. 1983*a*; Price *et al*. 1984; Tarin 1985; Schirrmacher 1985).

Many of these experiments have been performed with the B16 mouse melanoma, the Lewis lung carcinoma, or the Walker 256 rat carcinoma, all of which have now been propagated for many years, but later experiments with isogeneic rat carcinomas of recent origin (Murphy *et al*. 1988) have confirmed and strengthened the general conclusion that the distribution of metastases depends very much on the suitability of the new environment, and not just on anatomical factors. The results reported by Murphy *et al*. may be summarized as follows:

1. After the injection of tumour cells to a systemic vein or the portal vein, many were arrested in the 'organ of first encounter'. Cells reached bone by retrograde flow if the abdomen, and hence the inferior vena cava, was firmly compressed during injection of cells to the femoral vein.

2. The distribution of tumour *cells* injected into the left ventricle was similar to that of glass spheres (16 micron diam.) injected in the same way.

3. The distribution of tumour *colonies* after intraventricular injection of

tumour cells did not correspond to the sites at which cells were first arrested. The adrenal glands, ovaries and bone appeared to favour all the tumours tested; the relative numbers of colonies in the kidneys, brown fat, and, to a lesser extent, the lungs, differed for different tumours. Bioassay showed that tumour cells in the spleen and gut survived for between 24 hours and 5 days, and never gave rise to colonies; whereas in the kidneys cells remained viable but dormant until the animal died of tumour elsewhere.

4. Trauma, in the form of manipulation of the liver or a surgical wound, greatly favoured the development of colonies. This could not be attributed to a change in the blood supply because similar trauma had no significant effect on the distribution of glass spheres injected into the left ventricle. The authors have postulated that the effect of the trauma on the distribution of tumour colonies is due to growth factors released by macrophages at the site of injury.

The reasons why so many tumour cells are destroyed as a consequence of their 'first organ encounter' have not been fully elucidated. Weiss and Glaves (1983) have suggested that mechanical injury to the tumour cells, destruction by NK cells, and contact with vascular endothelium may all play a role. The preferential growth of tumour cells in a particular organ cannot be explained on immunological grounds because the distribution in athymic mice, and in rats made immunodeficient by administration of cyclosporin A, has been found to be much the same as in the corresponding normal animals (see Alexander 1985).

There have been some interesting studies which compare the ability of single tumour cells and cell clumps to form colonies and metastases.

It was observed by Liotta et al. (1974, 1976) that a transplanted murine fibrosarcoma, which regularly metastasizes, released both single cells and cell clumps into the venous effluent, and that the proportion of clumps of a given size was a linear function of the proportion of vessels large enough to contain them. This, together with the further observation that intravenous injection of aggregates of tumour cells was more likely to give rise to colonies in the lungs than intravenous injection of the same number of disaggregated cells, suggests that metastases are more likely to develop from cell clumps than from individual cells.

Despite this, there is evidence that a metastasis (Talmadge et al. 1982, 1984), or a lung colony developing after intravenous injection of tumour cells (Poste et al. 1982), may arise from a single cell, whereas different metastases (or colonies) develop from different cells. In the experiments of Talmadge et al., cells of a murine melanoma line were irradiated (650 rad) in vitro to induce chromosome breaks and rearrangements, and then injected into the footpads of compatible mice. Spontaneous lung metastases were isolated from different animals, established in culture as individual lines, and then karyotyped. With some metastases the same chromosomal abnormalities were present in all the cells examined, but most metastases differed from each other in that they exhibited characteristic combinations of chromosomal lesions. In so far as the metastases resulted from circulating clumps of cells, it seems clear, as Talmadge *et al.* point out, that either (1) each metastasis developed from just one cell of a heterogeneous clump, or (2) each clump was homogeneous because it originated from a 'clonal zone' of the primary neoplasm.

Cells in a clump have the advantage that they may be stimulated by growth factors secreted by neighbouring cells. It may be, as Alexander (1985) has speculated, that the development of colonies or metastases from a single cell depends on stimulation by appropriate growth factors provided by the surrounding normal cells.

5.2.3 Metastasis by surface implantation

Cancer cells implant readily on the endothelial surface of the pleura and peritoneum, and this typically results in an effusion of fluid containing tumour cells. In the case of pleural metastases the primary tumour is usually in the lung, whereas peritoneal metastases are most commonly derived from a primary in the ovary, stomach or colon.

Implantation may also occur on epithelial surfaces in the urinary and gastrointestinal tracts, and in the walls of viscera or the subcutaneous tissue when these structures are incised to permit excision of a primary tumour.

5.3 Is there a metastatic phenotype?

There is a sense in which it is a tautology to say that there is a metastatic phenotype: the fact that some cells metastasize implies that they have the ability to metastasize. But it is pertinent to ask whether metastases are derived from random survivors of disseminated cancer cells or, as Leighton (1965) first suggested, from a subpopulation of variant cells with heritable properties which are associated with a high metastatic capacity. In other words (Alexander 1982), is metastasis a *stochastic* or a *selective* process?

5.3.1 Is metastasis stochastic or selective?

The selective hypothesis is suggested by the fact that, as we have seen (Section 1.3.1), clones derived from the same tumour may differ markedly in their capacity to give rise to tumour colonies in the lungs after intravenous injection, and to metastasize after subcutaneous transplantation. This has been interpreted as evidence (Fidler and Kripke 1977) that the cells which give rise to metastases are members of a subpopulation that remains stable during cloning and clonal growth, but it now appears from studies with a wide variety of tumours that this is often not the case (Ling *et al.* 1985; Volpe and Milas 1988; Korycka and Hill 1989). With one mouse tumour there is evidence that the metastatic capacity of different clones depends on the level of expression of Class 1 MHC molecules, increased expression of H-2 K being associated with reduced metastatic capacity (rev. Feldman and Eisenbach 1988).

It is not possible to study the metastatic capacity of tumour cell subpopulations in humans but it seems likely that, if the selective hypothesis were correct, a primary tumour and its metastases would differ also in respect of other properties which could be observed. Some phenotypic differences have been reported from time to time, but so far no consistent pattern has been demonstrated (Weiss 1980*b*).

Many attempts have been made to resolve the question in animals by comparing cells from a primary tumour with cells from a lung colony produced by intravenous injection of tumour cells for their capacity to form lung colonies, or, what is better, cells from a primary tumour and cells from a metastasis for their capacity to metastasize.

Fidler (1973), in what appears to have been the first study of this kind, injected mice intravenously with cells of a cultured line of the B16 melanoma, harvested cells from a lung colony and established a subline from it, injected cells of the subline intravenously to another mouse, harvested cells from a lung colony and established a second subline, and so on. He found that the number of lung colonies that developed after intravenous injection of a standard number of cells was greater for each successive subline, and concluded that the 'survival of circulating tumor emboli is not a random phenomenon'. In later experiments Talmadge and Fidler (1982) confirmed that, starting with a subline of the B16 melanoma that produced relatively few lung colonies when injected intravenously, it was possible to select sublines of increasing colony forming capacity, and further, by starting with another subline that produced few metastases from subcutaneous transplants, sublines of increasing metastatic capacity could be selected. In both cases, however, after some generations, sublines were obtained which had reached a peak, and their performance could not be improved by further attempts at selection.

Giavazzi *et al.* (1980), in experiments with a variety of mouse tumours, including, in addition to the B16 melanoma, two chemically induced fibrosarcomas of recent origin, a colon carcinoma and an ovarian carcinoma, found that some metastases or lung colonies differed from each other, or from the primary transplant, in respect of their metastatic or colony forming capacity. This was unusual, however, and, as a general rule, cells from a lung colony were no more capable of forming colonies, nor were cells from a metastasis more metastatic, than those of the primary tumour. Furthermore, cells from a metastasis in a particular site did not show any tendency to 'home' to the same site in a later generation.

Alexander (1982) also found no evidence in favour of selective metastasis in experiments with rat sarcomas of recent origin, while Weiss et al. (1983b) found that with some tumours (including the B16 melanoma) cells from a metastasis showed an increased metastatic capacity, but that the opposite was the case with others.

Vaage (1988) compared the metastatic capacity of serially transplanted tissue from autochthonous primary mouse mammary carcinomas and their metastases, starting with seven mice (six C3H/He mice carrying the mouse mammary tumour virus and one C3Hf/He that was free of the virus), each of which had a single primary mammary carcinoma and one or more metastases. In no cases was any significant difference observed. Changes in metastatic capacity sometimes occurred during serial transplantation, but always in about the same transplant generation in the line set up with metastatic tissue as in the parallel line set up with the corresponding primary tumour. In later experiments, Vaage (1989) observed that, among 25 C3H/He and C3Hf/He mammary carcinomas of spontaneous origin that had all produced pulmonary metastases in autochthonous mice and also during subsequent intramammary syngeneic passage, the change in three cases took the form of loss of metastatic capacity. When these tumours were restarted in serial intramammary passage from early cryopreserved transplant generations, loss of metastatic capacity again occurred in the same (or nearly the same) transplant generation as before. Vaage suggests that the loss of metastasizing capacity may have been determined by 'a programmed genetic event', and is not inconsistent with the hypothesis that metastasis is a random process.

The experiments we have been discussing provide little evidence in favour of a metastatic phenotype, apart from those performed with lines obtained by cloning *in vitro*, often from tumours like the B16 melanoma that have been maintained in culture for a long time, and which may not reflect the behaviour of autochthonous tumours. But failure to demonstrate the existence of a metastatic phenotype does not rule out this possibility because, as we have seen (Section 4.4.2, pp. 49–50), cloned tumour cell populations may rapidly diversify both *in vitro* and *in vivo*, and a metastasis might therefore soon cease to be representative of the cell or cells from which it arose.

5.3.2 Genetic basis of metastatic ability

Reference has been made (Section 5.2.1) to the influence of Class 1 MHC molecules on the metastatic capacity of a mouse tumour. Other experiments with this tumour point to the existence of a gene not associated with MHC expression which also influences metastatic capacity, and it has been suggested that this codes for a receptor for a factor specific for growth of the tumour in lung tissue (rev. Feldman and Eisenbach 1988).

The genetic basis of the metastatic ability of human tumours is being investigated by testing tumour sublines of different metastatic ability for the expression of particular oncogenes, and by studying the effect of *in vitro* transfection on subsequent metastatic performance.

Schwab *et al.* (1984) reported that amplification of N-*myc* was a feature of advanced neuroblastoma metastases. More recently, Pohl *et al.* (1988) tested various related tumour cell sublines for oncogene expression. They observed small differences in the expression of several oncogenes between a lymphoma line of low metastatic ability (Eb) and a highly metastatic variant line (ESb). They also observed in one instance a 30-fold amplification of Ha-*ras* in a line derived from a metastasis of a spontaneous mouse mammary carcinoma, but no amplification in a line derived from the primary tumour. Somewhat surprisingly, the metastasis-derived line, despite the gene amplification, did not show increased expression of p21-Ha-*ras* protein.

Several groups have studied the effect of transfection with DNA from various sources on the tumorigenicity and metastatic ability of NIH/3T3 cells (Section 2.4.1, pp. 11–12) in nude Balb/c mice. Greig *et al.* (1985) found that NIH/3T3 cells transfected with c-Ha-*ras*-1 from the T24 human bladder carcinoma cell line gave rise to rapidly growing metastasizing tumours when injected to either a subcutaneous site in the supraclavicular region or to a footpad, whereas non-transfected cells gave rise to non-metastasizing tumours on supraclavicular injection and to slow-growing tumours that metastasized in only two of five mice after footpad injection. In similar experiments by Thorgeirsson *et al.* (1985) the transfected cells were both tumorigenic and metastatic on subcutaneous injection, whereas non-transfected cells were not tumorigenic except in the case of one 3T3 subline that had undergone spontaneous transformation, and this was not metastatic.

Bernstein and Weinberg (1985) found that fibroblasts of the NIH 3T3 line which had been transfected with the Ha-*ras* oncogene from the EJ human bladder carcinoma cell line and injected into normal mice formed non-metastasizing tumours, whereas after transfection with DNA from a human metastatic tumour the cells gave rise to a metastasis in one mouse, and cells transfected with DNA from this metastasis gave rise to metastases in several mice. DNA from all these metastases was shown to contain a particular fragment of human DNA, and it was concluded that 'the metastatic phenotype can be transferred from cell to cell and is associated with the presence of a discrete DNA segment'. This statement seems to imply that the DNA fragment in question belongs uniquely to metastatic tumour, but such a conclusion is not justified. To establish it, one would need to set up control experiments from the corresponding human primary tumour, and show that these yielded negative results.

Similar experiments have been reported by Radler-Pohl *et al.* (1988), who transfected cells of a mouse bladder carcinoma cell line with DNA from hepatic metastases of human colon carcinomas, and showed that these had a greater metastatic capacity than cells of the same line transfected with calf thymus DNA. Unfortunately, once again, no experiments were set up with cells transfected with DNA from the corresponding primary tumours, and without these it seems unjustified to draw any conclusions concerning the existence of a metastatic phenotype.

A possibility that must be considered is that different selective pressures operate at different stages in the 'metastatic cascade'. Evidence of this is provided by experiments of Vousden et al. (1986) in which mouse mammary carcinoma cells that were tumorigenic but non-metastatic were transfected with c-Ha-*ras* from the bladder carcinoma line. On syngeneic transplantation, the transfected cells metastasized more readily, and to more sites, than the control (non-transfected) cells, and all the metastases except one expressed the transfected gene; moreover, when cells from the metastasis that had lost the gene were re-transplanted, the incidence of metastases was no greater than in the controls. On the other hand, the incidence of lung colonies after intravenous injection of cells was unaffected by the presence of the *ras* gene. It was therefore concluded that the transfected gene increased the ability of the cells to escape from the primary tumour, but not their ability to survive in the circulation and seed to a new site.

5.4 Dormancy

5.4.1 Dormancy as a clinical phenomenon

Many years—in one recorded case as much as 50 years—may elapse between the apparently complete local ablation of a primary malignant neoplasm and the appearance of metastases (Everson and Cole 1966). This delay is particularly prone to occur with malignant melanoma, and with carcinomas of the breast, kidney and ovary.

During the time when metastases are present but not manifest, the tumour cell population must be relatively small. If this state persists for a long time the cell population must be stationary or growing very slowly, and the metastases are said to be *dormant* or *latent*. The diagnosis of dormancy must be *retrospective*, since a metastasis cannot be known to have been present until it has ceased to be dormant. It is also somewhat arbitrary, because there is no precise definition of what constitutes a long time in this context, and the size of the metastases that we can detect depends very much on their location, and on whether we rely on ordinary physical examination plus conventional radiography, or use more sophisticated scanning procedures (see Woodruff 1972).

Sometimes the appearance of metastases seems clearly to have been triggered by some definite event; often, however, no cause can be discerned. The following two cases illustrate these possibilities.

Case 1

This patient (Woodruff 1961, 1975), a female aged 50, presented with a lump in the left breast which proved to be a carcinoma. There were enlarged, firm lymph nodes in the axilla. Three years previously a malignant melanoma had been removed from her leg, and there had been no evidence of recurrence or metastasis. She was treated by radical mastectomy and, because many of the axillary nodes contained metastatic breast carcinoma, this was followed by a course of post-operative radiotherapy.

Six weeks after the mastectomy subcutaneous metastatic melanoma nodules appeared, first in the field of irradiation (Fig. 4) and then elsewhere. Some

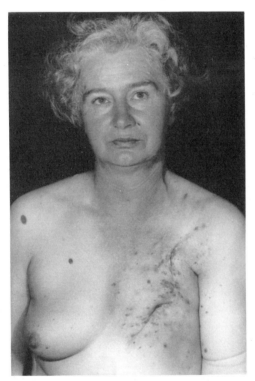

Fig. 4. Subcutaneous melanoma metastases which appeared after treatment of a mammary carcinoma by radical mastectomy and radiotherapy. A primary malignant melanoma on the leg had been removed three years previously. (By permission of Mr V. T. Pearse, FRCS and the University of Otago Medical School.)

4 months later the patient died, and at autopsy there were extensive melanoma metastases in the lungs, liver, bone marrow and brain.

It seems clear that the treatment, and in particular the irradiation, precipitated a change in the behaviour of disseminated but dormant melanoma cells; what is not clear is whether it did so by its effect on the melanoma cells, the local environment or both. There is no evidence concerning the underlying mechanisms, though it is plausible to speculate that, so far as the local environment is concerned, one factor might be impaired function of NC cells (see Woodruff 1986*b*).

In retrospect, it would have been of interest to have studied the karyotypes of cells from different metastases to see whether or not there was a consistent pattern of chromosomal damage in individual metastases, but unfortunately this was not thought of in 1957, when the patient died.

Case 2

This patient (Woodruff *et al.* 1976), a male aged 55, was referred to the writer with a huge carcinoma of the right kidney, and a history of having had a nephrectomy for a carcinoma of the left kidney four years previously. He was treated by right nephrectomy and transplantation of a cadaver kidney, and in the course of the nephrectomy it was observed that the tumour had invaded the renal vein. Despite this, and the fact that the patient received routine immunosuppressive therapy with azathioprine and prednisolone, he remained well without evidence of recurrence or metastasis for seven years. He then began to lose weight, and a radiograph of the chest raised a suspicion of pulmonary metastases. Within a few months he was dead, and at autopsy there were numerous metastases in the lungs and the liver. The sudden change in behaviour was not preceded by any change in treatment, or by any other event to which it could be attributed.

5.4.2 Experimental study of dormancy

Two procedures have been used to study dormancy in experimental animals: (1) injection of viable tumour cells into a systemic vein, the portal vein or the peritoneal cavity; (2) subcutaneous transplantation of tumours which commonly metastasize.

Dormancy after intravenous or intraperitoneal injection of tumour cells

It was reported by Fisher and Fisher (1959) that when rats were given an injection of 50 viable cells of the Walker rat carcinoma into the portal vein they normally failed to develop tumours during the next five months. If, however, starting three months after the injection they were subjected to repeated laparotomy and examination of the liver at weekly intervals, they all developed tumours within a few weeks.

More recently Wheelock *et al.* (1982) immunized mice with an intraperitoneal (ip) injection of mitomycin C-treated cells of a syngeneic lymphoma and challenged them later with an ip injection of viable cells of the same tumour. Some of the mice showed no evidence of tumour for many months and then suddenly developed ascites associated with intraperitoneal tumour. The period of dormancy was associated with a strong peritoneal cytotoxic T-lymphocyte reaction, and escape from dormancy with waning of this reaction and phenotypic changes in the tumour cells. Schirrmacher *et al.* (1982), in experiments with the same lymphoma, injected viable cells into the hind footpads of mice, where they failed to grow, and then challenged the animals with a subcutaneous injection of tumour cells in the back, where the tumour normally grows. It failed to do so in the mice that had previously received an injection in the footpad, but in 20 per cent of the challenged mice a tumour developed in the footpad. It was concluded that the cells injected to the footpad became dormant because they induced a state of immunity, and ceased to remain dormant as a result of immune stimulation resulting from the injection in the back. Dormancy could also be terminated by injecting allogeneic tumour cells subcutaneously in the back, showing that it was not necessary for the immune stimulation to be specific.

Dormancy of true metastases

Eccles and Alexander (1975) found that only 10 per cent of rats normally developed overt pulmonary metastases within 18 months of amputation of a limb bearing a subcutaneous transplant of a syngeneic fibrosarcoma. If, however, 1, 7, or 30 days after amputation the rats were exposed to 500 rad whole body irradiation, or were subjected to seven days continuous drainage of thoracic duct lymph—a procedure that is immunosuppressive and promotes the survival of allogeneic skin grafts (Woodruff and Anderson 1963)—there was a highly significant increase in the proportion which developed tumours during the 18 months period of observation.

5.4.3 Nature of the phenomenon. Some unanswered questions

In a dormant metastasis the neoplastic cell population remains small, and this implies either that the neoplastic cells are in G_O or cycling very slowly, or that the production of new cells is balanced by cell loss. In principle, this balance could be at any level of cell turnover. In practice, however, while in some established tumours, notably basal cell carcinomas of the skin, a high level of cell production may be approximately balanced for a considerable time by cell loss, this seems implausible as an explanation of dormant metastases.

Various reasons have been suggested to account for the low level of cell turnover, including inability of the tumour cells to develop a vascularized stroma, inadequate supply of necessary growth factors by the host, inability of the tumour cells themselves to produce sufficient growth factors to compensate for this, and the inhibitory effect of an immunological or para-immunological (Section 2.1) host reaction. Down regulation of the host reaction seems clearly to have been responsible for escape from dormancy in the experiments of Eccles and Alexander cited above. It also seems likely to have played a role in cases in which a tumour has developed in a kidney transplanted from a donor known to have a cancer elsewhere in the body (Penn 1970) but whose kidney appeared to be tumour-free, as a consequence of the immunosuppressive treatment given to the recipient. In one such case three years elapsed before a tumour became apparent in the transplant, and subsequently this 'allogeneic metastasis' went on to metastasize to the lungs in the recipient. Immunosuppressive therapy was stopped and one month later the transplant was removed. The pulmonary metastases continued to enlarge for about three months and then regressed completely. To reassure readers who may one day need an organ transplant it may be added that nowadays organs are not transplanted from donors with cancer.

What causes the sudden appearance of metastases and their rapid growth?

The first step might conceivably take one of two forms: (1) one or more cells in each of a number of small foci of dormant cells begin to proliferate because some environmental constraint is removed; (2) a heritable change occurs in one or more cells among a population of dormant cells distributed in small foci. Clearly, if only one cell undergoes such a change, there will be only one manifest metastasis until this itself metastasizes.

Following the initial step, the expanding tumour cell population may undergo progression, and some of the metastases that have appeared may, conceivably, themselves metastasize.

It would help us to distinguish between these theoretical possibilities if we knew whether the cells of each metastasis that appears, or even of the whole set of metastases, are the progeny of a single disseminated cell. This, as was mentioned in discussing the first of the cases cited above, is open to investigation if the termination of dormancy has been precipitated by some factor such as irradiation that causes chromosomal damage, but no studies of this kind appear to have been reported.

Given reasonable estimates of the volume of a metastasis, it is possible to fit a growth curve, and various functions relating volume (V) and time (t) have been tried, most commonly an exponential function or a Gompertz function. Attempts have then been made to extrapolate the curve backwards to the point where V is equal to the volume of a single cell, and so deduce when the metastasis began. The reader is referred elsewhere (Woodruff 1980, pp. 62–6) for a discussion of the mathematical properties of these functions and of the dubious propriety of the extrapolation, but we must consider one point that is relevant to the present discussion. With cases like the two cited in Section 5.4.1, it seems virtually impossible to fit a single growth curve $V = f(t)$ for which the specific growth rate dV/dt is continuous, because once they appear the metastases grow so rapidly, and while they are dormant they must have been growing very slowly, if at all. This, however, is entirely consistent with the view that escape from dormancy is a step in progression, as Foulds defined this process.

5.5 Regression of metastases

The case described in Section 5.4.3 is an example, possibly unique, of regression of metastases from an allogeneic tumour in a patient. Regression of blood-borne *autochthonous* metastases also occurs, and Everson and Cole (1966) collected 106 examples, in which the primary tumours included carcinoma of the kidney (23 cases), neuroblastoma (19 cases), malignant melanoma (15 cases), chorion carcinoma (7 cases), carcinoma of the ovary (7 cases), and osteogenic sarcoma (3 cases). Usually, though by no means always, regression occurred after excision of the primary tumour, conceivably because this removed a major source of growth factors. As mentioned earlier (Section 2.6.2), metastases from testicular germ-line tumours sometimes assume the appearance of a benign teratoma.

A remarkable phenomenon was reported by Bodenham (1968), who, by taking serial photographs of patients with multiple subcutaneous melanoma metastases, showed that some of these metastases regress while others are growing, and new metastases are appearing in the vicinity (Fig. 5). This cannot be explained in terms of purely systemic factors such as the immunological reaction of the host or the availability of growth factors of host origin; we are driven, therefore, to postulate differences in either the local environment or the intrinsic properties of the tumour cells. It would clearly be of interest in future cases showing the same phenomenon to excise adjacent progressing and regressing nodules for examination. Even studying their histological structure might reveal interesting differences, and this would pave the way for deeper investigation at the molecular level.

The extent to which metastases in lymph nodes regress is far from clear. There is no doubt that palpably enlarged nodes in the region draining a primary tumour sometimes cease to be palpable after removal of the primary, as for example axillary lymph nodes after simple mastectomy or partial mastectomy in patients with carcinoma of the breast, but, in an undetermined proportion of cases, enlargement of the nodes may have been a manifestation of the host reaction to the tumour and not the result of metastasis. In the days when most patients with breast cancer were treated by radical (or modified radical) mastectomy, attempts were made to gain some indication of the proportion of patients in whom enlargement of axillary nodes was due to metastases by histological examination of the nodes in the block of tissue excised at the operation, but unfortunately the figure arrived at depended very much on the thoroughness of the pathological examination; indeed, a distinguished pathologist once claimed that he could *always* find cancer foci in these specimens if he spent sufficient time and looked hard enough (Green 1950). The fact that some patients have remained tumour-free for many years after so-called radical mastectomy, which removes the axillary nodes but not the nodes in the anterior mediastinum alongside the internal mammary artery, without any other treatment, shows, however, that at the very least lymph node metastases may remain dormant for a long time, and suggests that they sometimes regress completely.

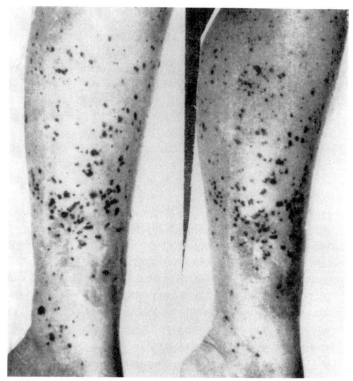

Fig. 5. Photographs taken at an interval of 12 months in a patient with smouldering disseminated malignant melanoma. Some lesions have faded, other ones have appeared (female aged 60). (From Bodenham, D. C. (1968). *Ann. R. Coll. Surg.*, **43.** By permission of the Royal College of Surgeons of England and the author.)

6
Limitations of current treatment

6.1 Assessing the results of treatment

Assessments of the effectiveness of treatment in cancer patients are commonly based on three parameters:

(1) how long the patient survives;

(2) how long he (she) remains clinically cancer free;

(3) the quality of life he (she) enjoys.

Length of survival is clear cut, but is influenced by various factors unrelated to the cancer or the treatment. Survival rates adjusted for age and sex are of much greater significance than crude survival rates, but leave out of account the other factors involved. The disease-free interval is less sharply defined than the survival time because it depends on the thoroughness of the search for residual cancer; and the question arises of how to deal with patients left after treatment with enlarged lymph nodes which may or may not be the site of metastases, and which remain stationary in size for varying lengths of time. Quality of life is still more difficult to quantify but various indices have been devised that enable useful comparisons to be made (see Karnofsky *et al.* 1948; Little 1987).

Randomized clinical trials are widely used to compare the benefits and complications of different forms of treatment. The design and analysis of such trials, their limitations, and the ethical problems they entail have been discussed in many publications (see, for example, Peto *et al.* 1976, 1977; Woodruff 1980, pp. 4-7; Cuzick *et al.* 1987*a*, 1987*b*), and will not be considered here.

Is cancer curable? Georges Mathé (1977) recounts how Général de Gaulle once asked him this question and he replied with another: 'Guérit-on jamais de cette drôle de maladie qu'est la vie?' This does not really answer de Gaulle, but it serves to remind us that we need to consider carefully what we mean by *cure*.

The best possible result of treatment for any neoplastic condition would be to destroy all the descendants of the founder cell (or cells) without causing significant harm to the patient, but for clinical use we need a less exacting criterion. One possibility is to regard a patient as cured if the lesion becomes and remains undetectable, and the patient lives as long as would have been expected if the neoplasm had never existed. We can never be certain, while a patient is alive, that he or she has been cured in this sense, and even at autopsy the presence of some cancer cells can never be completely excluded. If, however, we are considering a population of patients with cancer of the same kind treated in the same way, and neglect the seemingly remote possibility that recurrence may occur without affecting survival, the proportion of patients surviving when the survival curve for the patients first becomes parallel to that of a control population matched for age and sex is a plausible estimate of the cure rate for the population (Haybittle 1964), because at that time the risk of death in the patient and control populations is the same.

6.2 Local treatment

Local treatment in the form of surgical procedures or local radiotherapy aims to destroy a primary tumour, and sometimes also associated lymph node metastases.

Surgical treatment, which includes surgical excision and ablation of a tumour by electrocoagulation or with a laser beam, still plays a major role in the treatment of cancer and seems likely to continue to do so for quite a long time. It suffices to cure many patients with skin cancer, including virtually all patients with basal cell carcinomas who present reasonably early. It also offers good prospects for prolonged survival without recurrence of the tumour in many patients with cancer of the breast, cervix, colon and other sites.

Local irradiation, under which heading is included irradiation from an external source and implantation of radioactive materials directly into a tumour, but not systemic administration of radioactive substances (which is included under chemotherapy), offers an alternative to surgery in patients with radiosensitive tumours. It can be less mutilating than surgery, but this is by no means always the case, and if the tumour recurs further radiotherapy is usually impossible. Radiotherapy is often preferred to surgery, particularly in the United Kingdom, for the treatment of primary cancer of the lip, tongue, buccal cavity, larynx and cervix, and is quite often used together with surgery in the treatment of carcinoma of the breast. Radiotherapy is also used palliatively, especially to relieve pain in patients with metastases in bone.

6.2.1 Limitations of local treatment

Local treatment is subject to two limitations. First, if the primary tumour involves some vital structure, it may be impossible to eradicate it without causing unacceptable damage to normal tissue, in which case the tumour is said to be *locally untreatable*. Secondly, if there are distant metastases, local treatment does not attack these directly, and although metastases occasionally regress spontaneously after removal of the primary tumour (Section 5.5), the chance of this happening is very small. Conceivably, radical surgery which necessitates blood transfusion may actually promote the growth of existing metastases. There is no *decisive* evidence of this in man (rev. Woodruff and van Rood 1983) but the possibility is raised by the fact that blood transfusion may prolong the survival of organ allografts.

Advances in surgical technique, including the use of lasers and the possibility of replacing whole organs by transplantation, have resulted in more tumours being locally operable, but it is often not possible to take advantage of this because of distant metastases.

The scope of radiotherapy has been greatly increased by the introduction of linear accelerators and other machines that deliver high-energy X-rays. Implantation of radioactive material, which began many years ago with the use of radium needles and continued with the development of radon seeds and radioactive wire of various kinds, virtually disappeared when these machines became widely available, but seems to be coming back into use with the introduction of removable ^{192}Iridium wire, and seeds containing ^{125}Iodine that can be left in place indefinitely (see Ash 1986). It enables a high dose of radiation to be delivered to the tumour, with a steep fall off that spares neighbouring normal tissue, and may have a useful role in the treatment of cancer of the bladder and prostate; it seems unlikely, however, to represent a major advance in radiotherapy.

Two other developments in radiotherapy may be mentioned briefly, though both have been disappointing.

The first is the use of hyperbaric oxygen or chemical radiosensitizing agents to counteract the effect of hypoxia in tumour cells which tends to make them radioresistant (Adams *et al.* 1976; Williams 1985). One report on hyperbaric oxygen appeared encouraging (Henk and Smith 1977), but this promise has not been fulfilled. Chemical radiosensitizing agents have also proved disappointing. Mitchell (1971) has recorded that he and his colleagues tested 340 compounds, of which a synthetic vitamin K substitute (Synkavit) was the most effective. Except in one trial, however, the clinical results were described as 'modest or negative'. More recently misonidazole has been used as a radiosensitizer, but its main use is in multi-drug chemotherapy by virtue of its cytotoxicity for hypoxic cells, which in addition to being radioresistant are insensitive to many other cytotoxic drugs (Section 7.1).

The other is the use of a beam of fast neutrons instead of photons (gamma rays or X-rays). This also might be expected to be advantageous under hypoxic conditions, and first reports were encouraging (Caterall *et al.* 1977), but once again the early promise has not been fulfilled (Duncan *et al.* 1985; Battermann and Mijnheer 1986).

To what extent can the limitations of local treatment be overcome by earlier diagnosis?

If cancer originates at one site, and from one cell or just a few adjacent cells, there must in principle be some period after the first cell has taken the first step towards transformation during which the neoplasm is sufficiently localized to be cured by purely local treatment, though this may be vanishingly small in the case of the leukaemias, and with many solid tumours much shorter than was at one time thought to be the case. The probability (p_C) that local treatment will suffice to eliminate the cancer completely decreases with time, except in the case of tumours which grow very slowly and do not metastasize; the probability (p_D) of being able to diagnose the condition clinically is initially zero and increases with time. The general form of $p_C = F(t)$ and $p_D = f(t)$, expressed as functions of the time (t) that has elapsed since the beginning of carcinogenesis, is illustrated in Fig. 6. The stepwise nature of carcinogenesis

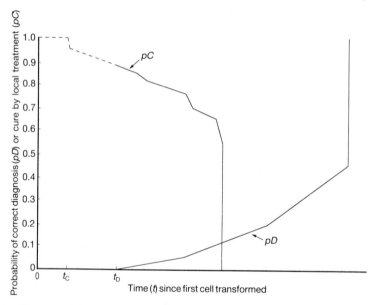

Fig. 6. Theoretical forms of curves expressing the probability of cure if the diagnosis were known (pC), and the probability of making a correct diagnosis (pD), as functions of the time (t) since the tumour began (see text).

and progression implies that $dF(t)/dt$ and $df(t)/dt$ will be discontinuous in places. When the risk of developing a tumour in a particular structure is very high, e.g. in a patient with an intra-abdominal testis, and in some patients with congenital adenopolyposis of the colon (Section 2.4.2, p. 22), it may sometimes be justifiable to perform a prophylactic operation to remove the structure at risk. Clearly, however, treatment (as distinct from prophylaxis) cannot be undertaken until there is a recognizable lesion to treat, and it is therefore of critical importance to know the probability of diagnosing the condition before dissemination has occurred. This probability differs greatly for different kinds of cancer, and for any given kind may be influenced by the alertness of the doctors of first referral, the diagnostic methods available, and whether or not patients at risk are subjected to regular screening. We shall illustrate this by considering three common forms of cancer.

Cutaneous malignant melanoma. The possibility of diagnosing this condition in many patients before dissemination has occurred is well shown by results reported from Queensland (Australia), where the incidence of malignant melanoma is very high. According to Davis *et al.* (1976), the overall age-adjusted five-year survival in more than 1100 cases was 81.6 per cent. Four years earlier the corresponding figure reported from Brisbane was only 69 per cent (McLeod *et al.* 1971), and very similar results have been reported from other parts of the world. Not surprisingly, Davis' results have not gone unchallenged, and it has been suggested that in some of his cases the lesion was benign, but independent examination of the histological sections has not supported this contention. It seems reasonable to conclude, therefore, as Davis *et al.* suggest, that the difference is due to earlier diagnosis, made possible by a vigorous campaign to alert both doctors and the public in Queensland to the potential danger of 'black spots' in the skin. To set up mass screening by doctors for malignant melanoma would be expensive, not because any expensive equipment is needed but because, to be effective, screening would require frequent attendance by almost everyone living in places where the risk is high, certainly several visits per year. In Queensland, at least, self-screening by members of the public as the first procedure seems to be proving remarkably effective.

Carcinoma of the cervix uteri. The discovery that carcinoma of the cervix can be detected by exfoliative cytology even at the stage of *carcinoma in situ* paved the way for the screening, initially of women with gynaecological complaints and later of 'well women', for this form of cancer. The results have been reviewed by Ellman (1986). Reports from many countries, including the Scandinavian countries, the United States and Holland, show a clear inverse correlation between the intensity of screening and the incidence of, and mortality from, invasive cervical cancer. In the United Kingdom no very clear evidence of the benefit of screening has emerged, but according to Ellman this can be attributed to large swings in the incidence of the disease, prompt and effective treatment when symptoms appear, and the gradual way in which screening was introduced.

Carcinoma of the breast. Randomized controlled trials in Sweden, the United States and the United Kingdom (rev. Ellman 1987) of the effect of screening by mammography strongly suggest that, provided the mammograms are of good quality and are reported on by radiologists trained to interpret them, the overall mortality from breast cancer can be reduced though far from abolished. In the Swedish trial reported by Tabar *et al.* (1985), for example, by the seventh year of the follow-up the mortality from breast cancer in women in the screened population who were aged 50 or more when they entered the trial was reduced by about a third, though no benefit was observed in younger women. Very similar results were reported earlier from the United States by Shapiro *et al.* (1982). It is noteworthy that the benefit of screening was still observed if, in patients with minimal or no evidence of cancer other than that provided by the mammography, the treatment was limited to excision of the lesion with a margin of healthy tissue. These findings suggest that in some, though far from all, patients with cancer of the breast the mammogram becomes positive before metastasis has occurred.

6.2.2 Local treatment of breast cancer

To illustrate the principles we have been discussing we shall look at the results of randomized trials of various forms of local treatment of breast cancer in the United Kingdom, other European countries and the United States, which have been recently reviewed by Cuzick *et al.* (1987a,b). The trials we shall consider were designed to make the following comparisons:

(1) radical mastectomy alone versus radical mastectomy plus postoperative radiotherapy;

(2) simple mastectomy alone versus simple mastectomy plus postoperative radiotherapy;

(3) radical mastectomy alone versus simple mastectomy plus radiotherapy.

Here *radical mastectomy* means either what is called Halsted's operation, in which the breast, both pectoral muscles, and the lymph nodes and fatty tissue of the axilla are removed *en bloc*, or a modified Halsted's operation in which the pectoralis major muscle is conserved. *Simple mastectomy* means removal of the breast, and sometimes also sampling of accessible axillary nodes for histological examination. The radiotherapy regimens differed in different trials in various ways, including the energy of the radiation, the extent of the field irradiated, the range of dosage and the duration of treatment. In all cases where irradiation was used the field included the axilla, and except in a small trial comprising 6 per cent of the total cases the dose was described as 'well above the minimum value estimated as adequate for local control in several reports in the literature'.

The only subgroups examined in the reviews were those based on age (over

50 or not), and on the status of the axillary lymph nodes (+ or −) assessed by clinical examination before randomization.

The total number of patients admitted to the trials was more than 11 000, and of these very nearly 50 per cent were still alive when last reported. When admitted all the patients exhibited what, at the time when the trials began, were termed Stage I or Stage II tumours; this means, in essence, that they were well within the limits of being locally treatable and may, or may not, have had palpable lymph nodes in the axilla.

Some *life tables* (for definition of this term see Peto *et al.* 1976, 1977) from the trials, not age-adjusted, are reproduced in Figs 7 and 8. It may come as

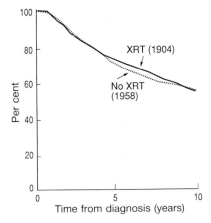

Fig. 7. Actuarial survival curves (life tables) for patients with operable breast cancer treated by simple mastectomy with or without postoperative radiotherapy. The numbers in parenthesis show the numbers of patients at risk in each arm of the trials from which the data were obtained. (From Cuzick et al. (1987b). *Cancer Treatment Reps.*, **71**. By permission of Dr J. Cuzick.)

a surprise to readers not involved in the clinical management of patients with breast cancer to see that both curves in both figures show 10-year actuarial survival rates of over 50 per cent.

The findings do not show any significant difference in survival during the first 10 years of follow-up between patients treated by radical mastectomy alone and those treated by simple mastectomy plus radiotherapy, nor between those treated by simple mastectomy plus radiotherapy and those treated by simple mastectomy alone. There is a suggestion that patients who are irradiated may do worse after 15 years' follow-up but the evidence for this falls far short of usually accepted levels of statistical significance.

Cuzick *et al.* (1987b) state that 'there is now little doubt that postoperative radiotherapy delays local recurrence', and cite references to support this, but they do not discuss what is meant by local recurrence. The writer knows of

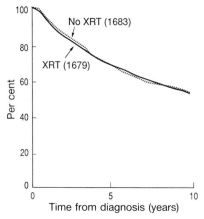

Fig. 8. Actuarial survival curves (life tables) for patients with operable breast cancer treated by radical mastectomy with or without postoperative radiotherapy. The numbers in parenthesis show the numbers of patients at risk in each arm of the trials from which the data were obtained. (From Cuzick et al. (1987b). *Cancer Treatment Reps.*, **71**. By permission of Dr J. Cuzick.)

one trial where node-positive patients who were treated by simple mastectomy only were classed as having local recurrence if the nodes simply remained clinically unchanged, a state of affairs for which *persistence* would seem a more appropriate term than *recurrence*.

Since there is no evidence that there is anything to be gained, in patients conforming to the conditions laid down for entry to these trials, by extending local treatment beyond simple mastectomy, it is pertinent to ask whether some lesser procedure might achieve equally long survival, at least in a subset of such patients, with less mutilation. Various trials designed to answer this question are in progress, and some have run long enough to point to an affirmative answer (see, for example, Hayward and Rubens 1987), but much remains to be done to define the optimal procedure in particular clinical situations. This work is important for patients with locally operable breast cancer and those responsible for their care; from the point of view of this book, however, the important conclusion to be drawn from the trials we have been discussing is that the main hope of significantly reducing the mortality from breast cancer lies in finding ways of preventing, permanently immobilizing, or destroying distant metastases.

6.2.3 Local treatment of other tumours

With many tumours, e.g. malignant melanoma and osteosarcoma, the possibility of total eradication by local treatment is most often limited, as in the case of carcinoma of the breast, by the presence of distal metastases. Other tumours, e.g. malignant gliomas of the brain, rarely if ever metastasize but often become locally inoperable, and with others again either limitation may come into operation first.

It is not necessary for our purpose to discuss particular tumours in detail; the need in many forms of cancer for effective systemic therapy is clear. We shall, however, have to consider the general question of whether it is advantageous to use, in addition to systemic treatment, local treatment to remove or destroy as much as possible of the primary tumour, when complete local removal is impossible (Section 6.4.2, pp. 82–84).

6.3 Wide field irradiation

Although wide field irradiation has been used very successfully in Stage I Hodgkin's lymphoma (see e.g. Doreen *et al*. 1984), a study by the British National Lymphoma Investigation Group (Haybittle *et al*. 1985), based on 14 years' experience in over 2000 cases of Hodgkin's lymphoma, points to the conclusion that it is not the treatment of choice. Instead, it is claimed, restricted field radiotherapy is indicated in a subset of patients classified as having Stage IA or IIA disease (which implies that the disease appears to be confined to nodes on one side of the diaphragm and is not associated with fever, sweats or loss of weight), and that all other patients should receive chemotherapy.

Irradiation of half the body is used palliatively for the relief of pain, and sometimes, in association with chemotherapy, in attempts to cure patients with Ewing's sarcoma of bone, Kaposi's sarcoma, neuroblastoma and small cell carcinoma of the lung. The two halves of the body may be irradiated in succession and, if sufficient time is allowed in between, bone marrow transplantation may not be required.

Whole body irradiation, followed by transplantation of either the patient's own bone marrow harvested prior to treatment or allogeneic marrow from a donor, is used in the treatment of various leukaemias and metastatic solid tumours. Nowadays, some form of chemotherapy is nearly always used in conjunction with these procedures.

The problems associated with bone marrow transplantation will be discussed in Section 6.4.2, pp. 84–86.

6.4 Chemotherapy

6.4.1 Scope and limitations

Under the heading of chemotherapy we shall discuss the administration, systemically or by regional perfusion, of cytotoxic substances other than hormones (which will be considered with other endocrinological procedures in Section 6.5). In principle, a distinction may be drawn between cytotoxic agents that are *cytocidal*, i.e. kill cells, and those that are *cytostatic*, i.e. arrest the progress of cells through the mitotic cycle; in practice, however, we do not know for how long a cell can be arrested in cycle and survive. A recent development is the use of cell-targeted drugs, but discussion of this will be postponed till Chapter 7.

Chemotherapy may be combined with other forms of treatment. A special case, for which we shall reserve the term *adjuvant therapy* (Section 6.4.2, pp. 82–84), is the use of chemotherapy to destroy potentially lethal occult metastases which may exist after local treatment of the primary tumour.

The substances used in cancer chemotherapy include *alkylating agents*, which damage DNA; *antimetabolites*, which interfere with the synthesis of purine or pyrimidine bases; and a variety of other substances of natural or synthetic origin which act by blocking transcription, inhibiting mitosis or in other ways, sometimes indeed in more than one way.

Cancer chemotherapy dates from about the end of World War II, but for some years it offered at best a modest degree of palliation, and at worst simply added to the patient's misery. Today, chemotherapy alone can be curative in acute lymphoblastic leukaemia of children, Hodgkin's lymphoma, Burkitt's lymphoma and chorion carcinoma, and, in association with local treatment, in Wilm's tumour, and in germ cell tumours, which occur most commonly in the testis. It can also prolong the disease-free survival in patients with many different kinds of cancer. Despite these great achievements, however, chemotherapy is still subject to serious limitations:

1. The drugs currently available are not sufficiently selective. To eradicate a neoplasm it is necessary to eradicate the neoplastic stem cells from which it arises, and this may be impossible without causing serious, and sometimes unacceptable, damage to normal cells.

2. The neoplastic cells may be, or may become, resistant to the drug or drugs used (Sections 6.4.3 and 7.1).

3. Most of the drugs used act by interfering with cell replication and do not destroy non-cycling cells.

4. Many anti-cancer drugs must be classed as carcinogens. The risk of subsequent neoplastic transformation may therefore be increased in normal cells that are affected, but not destroyed, by the treatment.

6.4.2 The problems of selectivity. Minimizing complications

The cytotoxic drugs in current use all damage normal cells. The tissues most at risk are those with a high cell turnover, especially intestinal epithelium, hair follicle epithelium, lymphoid tissue and bone marrow. Drugs differ, however, in their effects on normal tissues; some, for example, are relatively more damaging to intestinal epithelium than to bone marrow and with others the reverse is the case.

As we shall see in Chapter 7, much effort is being made to develop new drugs that will exploit subtle metabolic differences between normal and neoplastic cells, and ways of directing drugs to particular target cells. Meanwhile, there are choices to be made by the clinician that may markedly affect the outcome in particular patients:

1. What is the best regimen for systemic chemotherapy?
2. Should drugs be given by regional perfusion?
3. Should the bulk of a locally incurable tumour be reduced by surgery or radiotherapy before or after starting chemotherapy?
4. What measures should be used to counteract damage to lymphoid tissue and bone marrow? In this chapter we will discuss bone marrow transplantation. A development which is under investigation is the use of haemopoietic growth factors; this will be considered in Section 7.5.

Optimal regimens for systemic treatment

Treatment may be much more effective and less damaging if, instead of one drug, two or more are used in combination, particularly if these act in different ways. Moreover, it may be possible by alternating drug combinations to maximize the effect on the tumour while minimizing cumulative damage to normal tissue.

It was suggested many years ago (Berenbaum 1969) that optimal regimens might be designed in the light of dose-response curves for the cytotoxicity of individual drugs for neoplastic cells *in vitro*. Theoretical models for the design of optimal regimens, based on principles which emerged from studies of fluctuations in the virus-sensitivity of bacteria undertaken many years ago by Luria and Delbruck (1943), have been developed by Skipper and his colleagues (Skipper 1983, 1985; Skipper and Schabel 1984), and by Goldie and Coldman (1984). In practice, however, the initial choice of drugs, and the dosage and pattern of administration, are still based largely on clinical experience with neoplasms of the kind to be treated, and are modified according to the patient's response to treatment.

Prolonged, slow administration of drugs by intravenous infusion may be advantageous in patients with neoplasms that have a long cell-cycle time and a low growth fraction (see Ebbs 1987).

Regional perfusion

It is possible to increase the concentration of drug reaching a tumour without an increase of the same magnitude in the concentration of the drug in the general circulation, by injecting the drug into the artery from which the tumour derives its main blood supply.

This principle has been used, though without notable success, in patients with metastatic cancer of the liver. Although the tumour emboli from which these metastases originate reach the liver via the portal vein, the metastases typically derive much of their blood supply from the hepatic artery. This vessel is, therefore, cannulated under radiological control and the drug is injected into the cannula. Ligation of the hepatic artery is sometimes undertaken after, or used as an alternative to, hepatic artery perfusion (rev. Gray 1984). Infusion of cytotoxic drugs into the portal vein has been tried as a prophylactic procedure in patients with colon cancer which was treated surgically and in whom there were no overt hepatic metastases (Taylor et al. 1985), but the benefits were modest and some of the complications quite serious.

Another procedure that has been used in patients with malignant melanoma and other tumours affecting a limb is to connect the main artery and vein to a pump-oxygenator, occlude the vessels proximally, apply a tourniquet to the root of the limb, and inject the drug at a controlled rate into the circuit. The temperature of the blood being circulated is also controlled (rev. Sutherland 1986). A refinement that has been tried is to use an antimetabolite drug in the circuit and inject an antidote systemically. The main value of the local perfusion seems to be as a palliative procedure in patients with multiple subcutaneous melanoma metastases.

Adjuvant chemotherapy

The action of most cytotoxic drugs appears to conform over a wide dose range to what is sometimes called *first order kinetics*; i.e. the proportion of cells killed depends on the dose of the drug but is independent of the number of cells present (see Schabel 1969). This has prompted the suggestion that chemotherapy might be more effective if a tumour was reduced in bulk surgically, or by radiotherapy, before chemotherapy was begun (Schabel 1975; Burchenal 1976; De Vita 1983).

It seemed likely that this approach might be particularly rewarding in treating patients with tumours which carry a considerable risk of delayed mortality from micrometastases that are present but undetected at the time of adequate local treatment of the primary tumour. The term *adjuvant chemotherapy* was coined to denote the use of chemotherapy in such cases.

The first large-scale randomized trials of adjuvant chemotherapy were set up in the United States in patients with Stage I or Stage II breast cancer. All the patients were subjected to radical mastectomy, and on the day of operation and for two days thereafter received either chemotherapy or an inert placebo. No benefit resulted from the chemotherapy, and one of the drugs tried caused quite severe complications (Fisher 1971), but in trials undertaken in Scandinavia (Nissen-Meyer 1982), short-term (six-day) adjuvant chemotherapy resulted in prolonged survival. In randomized trials of long-term adjuvant therapy in patients with similar tumours, using either a single drug (Fisher *et al.* 1986) or a combination of drugs (Bonadonna *et al.* 1986), the disease-free interval was prolonged in premenopausal, but not in postmenopausal, women but in the earlier reports there was no convincing evidence of prolonged survival. Longer follow-up and data from later trials have confirmed the prolonged disease-free interval in premenopausal women and have shown further that survival is also prolonged in some sub-categories of patient with Stage II tumours. There appears to be little or no overall benefit in postmenopausal women, but Lippman and Chabner (1986) concluded that there was a reduction in the five-year mortality in those who received multiple drugs.

These trials have led to modifications in treatment that have been of benefit to many patients, but they do not seem to have brought us much nearer to a solution of the problem of how to eliminate occult residual cancer. There are, moreover, some grounds for thinking that the benefit observed in premenopausal women may be due to suppression of ovarian function rather than to a direct cytotoxic effect on tumour cells (Padmanabhan *et al.* 1986).

In high-grade osteosarcoma of the limb bones treated by amputation and adjuvant chemotherapy the survival rates at two years and five years were reported by a Medical Research Council Working Party (1986) to be no different from those observed 15 years ago, when treatment was by amputation alone, but a fairly recent randomized trial (Link *et al.* 1986) has confirmed reports from earlier trials that the disease-free interval is significantly prolonged by postoperative chemotherapy, though it had not been running long enough to provide conclusive survival data.

Adjuvant chemotherapy has also been used in association with surgical resection in patients with colorectal cancer in the hope of reducing the incidence of recurrence and hepatic metastases, but the benefit has again been rather modest. Hellman *et al.* 1987, for example, reported that administration of the cytotoxic drug razoxane (ICRF 159) five days per week after operation did not prolong survival or reduce the incidence of recurrence in patients with tumours confined to the bowel (Dukes' grade A or B), but halved the proportion of patients who developed hepatic metastases (median follow-up five years), and delayed their appearance, in patients with spread to lymph nodes (Dukes' grade C tumours). The choice of drug in this trial was based on experiments (Le Serve and Hellman 1972) in which razoxane delayed or prevented the metastasis of a murine tumour without affecting the growth of the primary, apparently by interfering with angioneogenesis (Section 7.6).

Why has adjuvant chemotherapy not been more successful?

An important question that is sometimes forgotten concerns the extent to which failure, when it occurs, is due to local recurrence which might have been avoided if the local treatment had been more radical.

So far as failure to destroy occult metastases is concerned, the first thing to be said is that the early optimistic expectations were based, explicitly or implicitly, on predictions of the fate of homogeneous cell populations whose members would either be eliminated as the result of exposure to a cytotoxic drug or would continue to multiply at a constant rate. With such populations it is possible, given the initial number of cells, the dose-response curve showing the fraction (F) of surviving cells after administration of a single dose (D) of a cytotoxic drug, and the actual dose and frequency of administration of the drug, to predict the number of cells surviving at any given time, and when this becomes less than one the tumour can be regarded as having been completely eradicated (see Berenbaum 1968, 1969). The dose response curve may be based on observations *in vitro* or, if surviving cells are assumed to continue to grow at a constant rate, on observations *in vivo*. In the simplest model the dose-response curve is assumed to be logarithmic, i.e. of the form $F = \exp(-aD)$, where a is a constant, and if this is modified to provide a 'shoulder' to the curve at low dosage by writing instead $F = 1 - [1 - \exp(-aD)]^b$, where a and b are constants, it fits well with observations with various alkylating agents (Skipper *et al.* 1964, 1965). With antimetabolites (Skipper *et al.* 1965), for reasons discussed by Berenbaum (1969), the dose response curve takes a different form, given by $F = [D/D_o]^{-c}$, where D_o represents the threshold dose of drug below which no cells are eliminated and c is another constant.

The validity of extrapolating dose response curves to the point where the population is reduced to a single cell is, to say the least, questionable, but these models are open to the further, decisive objection that they do not take into account the dynamic heterogeneity of tumour cell populations.

Two possible explanations of the origin of the metastases whose appearance marks the failure of adjuvant chemotherapy merit consideration:

1. Metastases develop from cells that were in G_o when the treatment was given and, in consequence, escaped destruction.

2. Metastases develop from a population of drug-resistant cells (Section 6.4.3) selected by the treatment.

Ragaz *et al.* (1985) have postulated that removal of a large part of the tumour cell population stimulates the remaining cells to proliferate, and under such conditions the chance of selection would be increased. In the light of their hypothesis they have set up a trial to examine the effect of giving a short course of preoperative chemotherapy to patients with early breast cancer after establishing the diagnosis by needle biopsy. At two years 86 per cent of the patients so treated were apparently tumour free and 96.7 per cent were still alive.

Bone marrow transplantation

Transplantation of autologous, isogeneic or allogeneic bone marrow can overcome in various degrees the damage to bone marrow caused by chemotherapy and whole body irradiation, and makes it possible under certain conditions to administer cytotoxic drugs in doses that would otherwise be lethal, and which may be sufficient to eradicate a neoplasm that would otherwise be incurable (rev. Thomas *et al*. 1975; Thomas 1984; Storb 1987). It appears moreover from experiments in animals (rev. Woodruff 1980, pp. 175–84, 241–2) that transplanted allogeneic marrow may have an antineoplastic effect, and there is evidence to suggest that this may be true also in patients.

Marrow transplantation has other uses but, so far as its role in neoplastic disease is concerned, it has been used mainly, but not exclusively, in patients with leukaemia. It currently provides the only hope of cure in chronic myeloid leukaemia (Storb 1987). An obvious, but important, consideration is that transplantation of marrow cannot make good the damage to other tissues such as the epithelium of the gastrointestinal tract, so that the effect on these tissues may be what limits the amount of chemotherapy that can be given.

Reinfusion of stored autologous marrow taken from a patient with leukaemia after inducing a remission, and from patients with other tumours before starting intensive chemotherapy, has the disadvantage that this may re-introduce a significant number of viable tumour cells and be responsible for recurrence of the disease. To avoid this, attempts are made to 'purge' the marrow of neoplastic cells either by immunological techniques involving the use of monoclonal antibodies or by treating the marrow with a cytotoxic drug (see, for example, Kushner *et al*. 1987) or a cytotoxic substance linked to a specific antibody (see Section 7.3.1). It is difficult to assess the efficacy of these procedures because ethical considerations limit the extent to which they can be tested in randomized clinical trials, and there are no genetic markers by which to distinguish between successful engraftment of autologous marrow and regeneration of marrow in the patient. It seems, however, that at present both techniques are associated with some increase in the risk that the graft will not become established.

Transplantation of marrow from a human donor other than an identical twin involves the risk of immunological rejection of the graft, and also the risk of graft-versus-host disease (GVHD).

Both risks can be reduced by a suitable choice of donor. Failing an identical twin, a healthy HLA-identical sibling is generally accepted as the best possible choice, but the degree of compatibility is much less than in the case of an identical twin donor and mismatches at the so-called minor histocompatibility loci, for which typing techniques have not yet been developed, may be responsible for severe GVHD.

Appropriately timed chemotherapy, with or without whole body irradiation, is used to attack the neoplasm, and to reduce both the rejection reaction of the host and the reaction by T-lymphocytes in the marrow that is responsible for GVHD. Another way to reduce the risk of GVHD is to remove as many as possible of these T-lymphocytes before transfusing the marrow. One technique that is used for this purpose is based on lectin agglutination and rosetting with sheep erythrocytes; another depends on monoclonal antibodies that react with particular T-cell subsets. Successful depletion of T-cells will have the effect of delaying recovery of the patient's normal resistance to infection, but this can be compensated for by strict isolation and the use of antibiotics when required.

6.4.3 Drug resistance

As we have seen (Section 1.3.1), the neoplastic cells of both experimental tumours, and spontaneously developing animal and human tumours, are markedly heterogeneous in respect of a wide variety of phenotypic characters, including sensitivity to cytotoxic drugs.

Tumour cells may be resistant to a particular drug *ab initio* or may become resistant during treatment as the result of selection (see, for example, Siracky 1979; Talmadge *et al.* 1984), and this plays a major role in limiting the effectiveness of cancer chemotherapy. It seemed reasonable to expect that the problem of drug resistance could be reduced, perhaps even abolished, by using combinations of drugs which act in different ways; it became clear, however, that the problem was much more difficult than it first seemed when it was discovered that exposure to one drug often results in increased resistance to a wide variety of drugs. This phenomenon, which is termed *pleiotropic resistance*, is commonly associated with the production of increased amounts of a plasma membrane glycoprotein (P-glycoprotein) which is thought to increase the efflux of certain classes of cytotoxic drug from the cell. The over-production of P-glycoprotein appears to be associated with amplification or increased transcription of a single gene, or a small number of closely linked genes (see Goldie and Coldman 1984; Bradley *et al.* 1988; Gottesman and Pasten 1988).

So far as its therapeutic consequences are concerned, the resistance of neoplastic cells to particular cytotoxic drugs resembles bacterial resistance to antibiotics, but the nature of the underlying changes in the cell are less well understood. The following possibilities merit consideration:

1. Changes in the cell membrane which retard the entry of drug to the cell, or accelerate its escape.

2. Changes in drug binding.

3. Changes in drug metabolism, including changes which alter the rate at which an inactive form of the drug is activated and *vice versa*.

4. Changes in the capacity of the cell to repair damage to DNA (Harris 1988). This seems likely to be important with drugs that interact with DNA directly.

The proportion of cells exhibiting the cytogenetic features of gene amplification (Section 4.2.2) may be increased by treatment with cytotoxic drugs, and gene amplification may be a major factor in the development of resistance to some categories of drug, especially the antimetabolites. Amplification of the gene coding for dihydrofolate reductase has been demonstrated in cells resistant to methotrexate, but increased transcription and translation may have similar effects.

The sensitivity of both normal and neoplastic cells to a particular drug varies during the cell cycle, and the phase at which sensitivity is maximal is not the same for all drugs (rev. H. O. Klein 1972). If the cycle time and the interval between successive doses of a drug are both long the therapeutic effect may be decidedly suboptimal. Cells that have stopped cycling temporarily (G_0 cells) appear to be undamaged by most of the cytotoxic drugs in common use or to be able to repair damaged DNA before re-entering the cycle.

6.4.4 The risk of inducing new neoplasms

To what extent, if any, do chemotherapy and radiotherapy increase the risk that a patient will develop a second cancer, or that his or her offspring will develop cancer.

The first question has been the subject of several recent investigations (Coleman *et al.* 1987; Hawkins *et al.* 1987; Kaldor *et al.* 1987; Kingston *et al.* 1987). The risk of a second primary cancer in patients who have been treated for various neoplastic conditions, including Hodgkin's disease, carcinoma of the ovary, testicular cancer, retinoblastoma, Wilms' tumour, osteosarcoma and acute leukaemia, is certainly greater than the risk of cancer developing in an age-matched population of people who have not had cancer previously. Much of the difference, however, is due to genetic predisposition, and environmental factors may also contribute to some extent, so it is difficult to estimate how much is due to the treatment of the original neoplasm (usually referred to in this context as the *index* neoplasm. The current view is that chemotherapy and radiotherapy may contribute significantly, especially when the dosage is high, but that the risk is small in relation to the benefit gained in the form of prolonged remission or cure of the index neoplasm, but for a detailed appraisal of the evidence the reader is referred to the references cited.

The risk of treatment-induced neoplasms in patients who receive adjuvant chemotherapy for breast cancer merits special consideration because of the limited benefits achieved so far.

In the American National Surgical Adjuvant Breast and Bowel Project (NSABP) trial, in which the patients were treated for two years with phenylalanine mustard, the cumulative risk of leukaemia and myeloproliferative syndrome was 1.7 per cent at 10 years as compared with 0.3 per cent for breast cancer patients in the trial who did not receive chemotherapy (Fisher *et al.* 1985). This difference is highly significant, but in the Milan trial reported by Bonadonna *et al.* (1986), in which three drugs were used in combination, the risk of leukaemia did not appear to be increased. In none of the trials reported so far has there been a significant increase in the incidence of solid tumours (see Lippman 1986), but longer follow-up is necessary before firm conclusions concerning the risk can be drawn.

The second question is *sub judice*. In a large retrospective study (Mulvihill *et al*. 1987) the incidence of cancer in the offspring of long-term survivors of childhood or adolescent cancer did not differ from that in sibling controls or in the general population, although there were differences in the distribution of different types of cancer. The follow-up was short, as the authors point out, but the findings suggest that risk of cancer in early childhood in the offspring of long-term survivors is small, except in the case of hereditary tumours. Even if overall differences in cancer incidence appear with longer follow-up, however, it seems likely that a still larger study would be needed to determine to what extent, if any, these can be attributed to treatment.

6.5 Endocrinological procedures

Endocrinological procedures used in treating cancer include the administration of hormones and hormone antagonists, and endocrine ablation by surgery or radiotherapy.

Prednisone is often used as one component of multiple drug chemotherapy in the treatment of acute lymphoblastic leukaemia and various other neoplasms. Apart from this, endocrinological procedures are used mainly in patients with carcinoma of the breast, prostate and thyroid, either for the treatment of advanced disease or as adjuvant therapy when there is a possibility of cure.

6.5.1 Treatment of advanced disease

Oophorectomy, or irradiation of the ovary in a dosage sufficient to abolish oestrogen production, is used in premenopausal women with advanced breast cancer, and is beneficial in about half the patients, especially in those whose tumours possess oestrogen receptors. If there is a good response followed by relapse, prevention of oestrogen production by the adrenal may produce a further remission. This was at one time achieved by adrenalectomy, hypophysectomy or abolition of pituitary endocrine function by implantation of ^{90}Ytrium into the pituitary fossa, but nowadays is achieved by administration of drugs such as aminoglutethamide. If there is no response to ovarian ablation there may nevertheless be a response to administration of the synthetic, non-steroid oestrogen antagonist *tamoxifen*, which competes for cytoplasmic oestrogen-binding protein. The tamoxifen-receptor complex is translocated to the nucleus, but this does not stimulate replenishment of the binding protein.

Postmenopausal patients with advanced carcinomas of the breast which possess oestrogen receptors may respond to tamoxifen, and also to large doses of the synthetic oestrogen *stilboestrol*; and the therapeutic effect may be increased if these are combined with chemotherapy (Kiang *et al.* 1985). The competition of tamoxifen for oestrogen receptors has also been demonstrated *in vitro* in studies with an oestrogen-sensitive human mammary carcinoma cell line (Lippman and Bolan 1975). The addition of oestradiol to the medium, up to a certain concentration, results in increased synthesis of DNA and protein, but this is prevented by simultaneous addition of tamoxifen. Since tamoxifen also depresses DNA synthesis below the control level in medium that is completely free of oestrogen, it seems that tamoxifen also has a direct cytotoxic effect on cells of this line. It has been shown in other experiments with the same line that high concentrations of oestradiol or stilboestrol depress DNA synthesis, whereas lower concentrations are stimulatory. In the case of stilboestrol, inhibition occurs at a concentration of 5×10^{-7} M, and this may explain the clinical antitumour effect of stilboestrol, since this level can easily be reached *in vivo*.

Orchidectomy, or administration of either an oestrogen or luteinizing hormone releasing hormone (LHRH), or both, is used in patients with disseminated carcinoma of the prostate, especially for the relief of symptoms such as bone pain. In many patients there is both subjective and objective improvement, though sometimes at the cost of unpleasant side effects. According to Williams (1988), however, there is no convincing evidence that survival is prolonged.

In patients with disseminated papillary or follicular carcinoma of the thyroid, a remission may be produced by administration of thyroxine in sufficiently large dosage to inhibit the production of pituitary thyrotrophic hormone, and in some cases by administration of ^{131}I. Alternation of these procedures may result in prolonged remission.

6.5.2 Endocrine adjuvant therapy

Thyroxine is administered routinely as adjuvant therapy after total or subtotal thyroidectomy in patients with operable papillary or follicular carcinoma of the thyroid.

The role of endocrine adjuvant therapy in operable breast cancer has been the subject of many clinical trials.

Initially, the procedure tested was ovarian ablation by surgical oophorectomy or ovarian irradiation. In many trials no benefit was observed; in others, survival was prolonged in premenopausal patients (rev. Lippman 1986). Some doctors who participated in other trials felt that trial of adjuvant oophorectomy was unjustified. Some patients, moreover, declined to participate in trials in which this was an option, and others withdrew if this option was allocated to them at randomization. It seems likely that more consistent results would be obtained if the oestrogen receptor (ER) status of patients was taken into consideration, but the prospect of a trial being set up to test this seems remote.

Current interest is centred on the oestrogen antagonist tamoxifen, which is remarkably well tolerated, and in 1987 there were more than 40 randomized trials of adjuvant tamoxifen in progress in the United States, and Europe, including the United Kingdom. The protocols, and sometimes the results obtained with similar protocols, differ considerably. The following summary of results is based on reports of long-running trials in Canada, the USA and Europe, presented at a Conference held in the USA in 1985 under the auspices of the National Cancer Institute (see Lippman 1986), together with three British trials reported elsewhere (Nolvadex Adjuvant Trial Organisation 1985; Ribeiro and Swindells 1985; Scottish Breast Cancer Trials Committee 1987).

1. Tamoxifen was well tolerated, even when given daily for five years.

2. In premenopausal patients there was, as a rule, little benefit. In two of the British trials, however, the disease-free interval was prolonged to the same extent as in postmenopausal patients, and in one of these patient survival was also prolonged.

3. In postmenopausal patients the disease-free interval was prolonged in those given tamoxifen in nearly all trials, and in some trials patient survival was also prolonged.

4. Administration of tamoxifen for five years, as in the Scottish trial, appeared to confer greater benefit than administration for one or two years.

5. In the Scottish trial, survival was longer when administration of tamoxifen was begun immediately after local treatment than when it was delayed until recurrence had occurred.

6. There is some disagreement concerning the value of the oestrogen receptor status of the tumour in predicting the response to tamoxifen. In many trials there was quite good correlation between the oestrogen receptor (ER) level and the response of the tumour to tamoxifen, and in consequence Lippman and Chabner (1986) concluded that it was not justified, even in postmenopausal patients, to use tamoxifen as the only adjuvant therapy if the tumour was ER negative. In two trials, however, the most that could be said was that only very high ER levels (more than 100 fmol/mg cytosol protein) were significant. McGuire *et al.* (1986), in a critical review of the literature, concluded that the presence or absence of progesterone receptors was a better predictor of the response to tamoxifen than the presence or absence of ER.

Until recently ER levels have usually been determined biochemically in cytosol preparations of tumour tissue (EORTC Breast Cancer Group 1980), and give no indication of how heterogeneous the tumour cells may be in this regard. It is now possible, however, by an immune cytochemical technique to determine the ER status of individual cells, and hence the proportion of cells bearing demonstrable ER, in samples obtained by needle biopsy of the primary tumour, recurrent tumour or soft-tissue metastases. It has been shown by using this procedure that many carcinomas of the breast contain both ER positive and ER negative cells (see, for example, McCarty *et al.* 1986), and this has prompted the suggestion that mammary carcinomas that are biochemically positive, but do not respond as expected to hormone therapy, fail to respond because they contain a sufficient number of ER negative cells (Jonat *et al.* 1986). This gains support from the observation of Coombs *et al.* (1987) that, in patients responding to tamoxifen, 58 per cent of the cells were ER positive, whereas in non-responders only 21 per cent of cells were positive. It seems, therefore, that we have here yet another example of how tumour cell heterogeneity influences the results of treatment.

7
Ways forward

7.1 Introduction

It seems unlikely that there is much, if anything, to be gained from more radical local treatment, though it is worth asking whether we are moving too far in the opposite direction. As was pointed out in Section 6.4.2, pp. 82–84, a critical analysis of the extent to which loco-regional recurrence is the first evidence of failure of systemic adjuvant therapy should help to answer this question.

If local treatment has reached its zenith, progress will depend on the development of more effective systemic treatment, to be used either alone or in combination with local treatment. Ideally, such systemic treatment should be, and should remain, cytotoxic to all the neoplastic cells in a given neoplasm but, in the dose needed to destroy the neoplasm, harmless to normal cells.

Empirical testing of vast numbers of natural or synthetic compounds for anti-cancer activity has proved unrewarding in the past (see Smyth 1980), and seems unlikely to be more cost-effective in the future.

A more promising approach is to use new combinations of existing agents to try to reduce the risk of drug resistance developing. Multidrug therapy is unlikely to prevent pleiotropic resistance (Section 6.4.3), but it may nevertheless be advantageous to use combined regimes which include drugs that attack hypoxic cells, which occur in poorly vascularized areas of solid tumours and are relatively insensitive to many of the anti-cancer drugs in common use (see Kennedy 1987; Sartorelli 1988). There are various reasons for this lack of sensitivity (Kennedy 1987): (1) some drugs, e.g. bleomycin, are activated only in the presence of oxygen; (2) hypoxic cells have prolonged cycle times or may even have stopped cycling; (3) DNA repair mechanisms may be impaired, and while this may favour cell death it involves the possibility that cells which survive may be more susceptible to mutagenesis, and in consequence more likely to become drug resistant. Misonidazole and mitomycin C (see Kennedy 1987) are examples of drugs that are toxic for hypoxic cells and/or increase their sensitivity to other agents, and the effectiveness of mitomycin C may be further increased in the presence of dicoumarin (Sartorelli 1988). The use of such drugs in combination with other drugs, or irradiation, which are preferentially toxic for well-oxygenated cells has proved advantageous in experiments with a murine mammary carcinoma, and some promising results have been reported in patients (Papac et al. 1987).

Exciting, more distant horizons are becoming discernible as the result of advances in molecular biology, nucleic acid chemistry, X-ray crystallography in relation to the determination of protein structure, computational methods and molecular modeling, and immunology, that have paved the way for the development of a variety of novel agents and procedures. These may conveniently be classified as follows, though there is some overlap between the categories:

1. Drugs designed to exploit metabolic differences between neoplastic and normal cells (see e.g. Harrap and Connors (ed.). 1987, Part I).

2. Drugs that induce differentiation in neoplastic cells (Section 2.6).

3. Drugs that modify gene expression in cancer cells (Section 7.2) or block the action of oncogene products (see Gullick and Sikora 1988).

4. Drugs that inhibit DNA repair mechanisms (Berger et al. 1987).

5. Conjugates of antibodies that recognize tumour-associated antigens and either (a) cytotoxic substances or (b) enzymes that can activate a systemically-administered pro-drug (Section 7.3).

6. Procedures that promote rejection of a neoplasm by the host (Section 7.4).

7. Administration of growth factors to counteract damage to normal tissues by cytotoxic drugs (Section 7.5).

8. Agents that inhibit the development of a vascularized stroma in solid tumours (Section 7.6).

9. Procedures to replace missing tumour suppressor genes.

The use of many of these agents falls under the heading of *biotherapy*. This term might be used to include the use of any substance of biological origin, but it is useful to restrict it (Oldham 1987) to the use of substances encoded by the mammalian genome, a growing number of which can now be manufactured by recombinant DNA technology. As Oldham has pointed out (*op. cit.*, p. 489), 'biotherapy [in contrast to chemotherapy] works through physiologic molecules to which the body has receptors and known mechanisms of activation'.

We shall examine the developments that seem to the writer most promising from the point of view of possible therapeutic application, though not necessarily in the immediate future. The discussion will be concerned with general principles. Readers in search of information about the technical aspects of anti-cancer drug design are referred to papers published in *Anti-Cancer Drug Design* and other journals devoted to this topic.

7.2 Control of gene expression

Intense efforts are being made to achieve greater selectivity for cancer cells, and avoid the risk of drug resistance developing as the result of gene amplification (Section 6.4.3), by developing agents that selectively regulate gene expression in tumour cells.

In principle, a therapeutic effect might be achieved either by up-regulating genes such as those coding for tumour associated antigens (TAA) or MHC Class I molecules, non-expression of which may favour survival of the tumour (see Woodruff 1980, pp. 86 *et seq.*), or by down-regulating genes that code for growth factors (GF) or GF receptors, and other oncogenes, or genes that confer drug resistance or contribute to cellular diversification. In practice, the problem of how to regulate the expression of particular genes in the desired direction is far from easy.

Agents such as azacytidine that inhibit DNA methylation have been shown to up-regulate various genes *in vitro*, including genes coding for TAA in a non-immunogenic variant of a normally immunogenic mouse lymphoma (Altevogt *et al.* 1986), but would be expected to act non-selectively on methylated sites in general.

Selective down-regulation, on the other hand, is now possible as a result of the development of agents that bind preferentially to particular base pair sequences in DNA and RNA. Since cellular proto-oncogenes may have important functions in normal cells, even in adults, the objective in down-regulation must be to suppress the expression of mutated oncogenes, or reduce appropriately the high expression of normal oncogenes in neoplastic cells (Miller and Ts'o 1987).

The sequence-specific agents that have been developed are of two main kinds: oligonucleotide analogues and oligopeptides.

7.2.1 Oligonucleotide analogues

Synthetic oligonucleotides can block translation of mRNA, but their value as therapeutic agents is limited by their poor penetration into cells and their sensitivity to destruction by intracellular nucleases (Thuong *et al.* 1987). Two kinds of oligonucleotide analogue have been developed which go far to overcome these limitations, namely, oligonucleoside methylphosphonates and oligonucleotides covalently linked to intercalating agents.

Oligonucleoside methylphosphonates

Miller and his colleagues have synthesized non-ionic oligonucleoside methylphosphonates (OMP) in which the negatively charged phosphodiester linkage normally found in nucleic acid is replaced by a non-charged phosphate group. These OMPs are more lipophilic than oligonucleotides, and thus better able to cross cell membranes; they are also more resistant to nucleases (Miller *et al*. 1981; Blake *et al*. 1985), but retain the capacity to form stable, hydrogen-bonded complexes with complementary sequences of single-stranded DNA and RNA. Their capacity to block gene expression depends on the fact that 'in order for a gene to be expressed, its nucleotide sequence must be exposed, at least transiently, in single-stranded form' (Miller and Ts'o 1987).

Miller and Ts'o (1987) refer to these OMP as *matagens*, this being an acronym for 'masking tape for gene expression'. They have used two strategies to block gene expression at the mRNA level. The first uses matagens complementary to specific sites such as the initiation codon to block translation of mRNA; the second uses matagens complementary to splice junctions of precursor mRNAs to inhibit splicing. A further development has been to attach reactive groups to the matagen molecule; one such group, aminomethyltrimethylpsoralen, results in a molecule that cross-links to the target nucleic acid only after irradiation with UV light of appropriate wavelength.

Oligonucleotides linked to intercalating agents

Helene and his colleagues (Asseline *et al*. 1984) have pioneered the use of oligonucleotides covalently linked to intercalating agents (which bind preferentially to double-stranded nucleic acid). The oligonucleotide retains its specificity for its complementary sequence, while the intercalating agent provides additional binding energy that stabilizes the complex. The covalent linkage at the 5' or 3' end of the oligonucleotide provides protection against exonucleases (but not endonucleases), and has been shown to facilitate uptake of the molecule by cells in tissue culture. It has been shown further that these modified oligonucleotides can block translation of mRNA in a specific way (Toulmé *et al*. 1986).

Resistance to nucleases can be markedly increased by using synthetic [α]-anomers of nucleosides as building blocks instead of the natural [β]-anomers. The resulting oligo-[α]-deoxynucleotides form double helices with complementary DNA sequences in which the two strands adopt a parallel orientation instead of the anti-parallel orientation characteristic of normal double-stranded DNA (Thuong *et al*. 1987; Helene 1988).

Both [α]- and [β]-oligonucleotides can be provided with 'teeth' by linking them covalently with reactive groups that can be activated either chemically or photochemically to generate irreversible reactions in their target sequences. Such agents can function, as Helene (1988) has graphically expressed it, as 'artificial nucleases', and open the way for the selective regulation of gene expression at the transcriptional level.

7.2.2 Oligopeptides

It has been suggested that preferential binding to particular base pair sequences in DNA, and consequential conformational changes extending some distance from the binding site, may account for the antiviral and antitumour activity of various naturally-occurring oligopeptides, e.g. distamycin and echinomycin, which are classed as antitumour antibiotics (Low et al. 1984; Fox and Waring 1984; Fox et al. 1988).

These agents bind, though non-covalently, to sites in the minor groove of double-stranded DNA; distamycin, for example, to sites consisting of five A-T base pairs. This has led Lown and his colleagues to synthesize information-reading oligopeptides which recognize and bind to other pre-determined sequences in the minor groove (Lown et al. 1986), which they call *lexitropsins*. They have gone on to synthesize oligopeptides bearing an alkylating moiety, and have reported that these show activity, in subtoxic concentrations, against some viruses, and animal and human tumour cell lines (Krowicki et al. 1988). It will be of interest to hear reports of the effect of these agents *in vivo*.

7.3 Antibody-directed attack on cancer cells

The development of monoclonal antibodies (mAbs) has rekindled interest in the possibility of antibody-directed attack on cancer. There have been some moderately encouraging results with antitumour antibodies (Ritz et al. 1981; Levy and Miller 1983), and also with anti-idiotype antibodies that can, under certain conditions, potentiate the host reaction (see Koprowski et al. 1984). The main thrust, however, has been to develop conjugates of antitumour mAbs with either (a) substances that are directly cytotoxic, including radionuclides, or (b) enzymes that can activate a systemically-administered pro-drug. Treatment with such conjugates is sometimes referred to as *immunochemotherapy*.

7.3.1 Immunochemotherapy

It should, in principle, be possible to destroy cancer cells selectively, while sparing normal cells, by administering intravenously, or into a serous cavity which is the site of a malignant effusion, a conjugate of an antibody that recognizes only the cancer cells and a chemical cytotoxin or radionuclide, if the following conditions are fulfilled:

1. The conjugate binds strongly, through its antibody, to antigenic determinants ('markers') on the surface of the cancer cells, but does not bind to normal cells.

2. The cytotoxic agent retains a high level of toxicity after conjugation with the antibody.

3. In the case of chemical cytotoxins the conjugate gains entry to the cell in sufficient concentration to be cytocidal.

In practice, it is not easy to make conjugates that come near to fulfilling these conditions. The reasons for this have been discussed in detail by Embleton (1987), and we shall give only a summary, based on his review.

Firstly, many of the mAbs that have been raised against human tumours cross-react to some extent with normal cells. This may be of little consequence if the antibody binds only weakly to normal cells, or only to cells that are not important as regards survival of the patient, but this is by no means always the case.

Secondly, the process of conjugation may greatly reduce the binding capacity of the antibody. This deleterious effect may be markedly reduced if the cytotoxic agent is first coupled to an inert protein, polypeptide or polysaccharide carrier, which is then attached on an equimolar basis to the antibody (Rowland *et al.* 1975; Hurwitz *et al.* 1978; Garnett and Baldwin 1986).

Thirdly, conjugates with chemical cytotoxins may be ineffective because the conjugate is only weakly cytotoxic or is unable to gain entry to the cell in adequate concentration.

The cytotoxins that have been conjugated with antibodies include drugs of the kind that are commonly used in unconjugated form, and plant toxins such as ricin.

The drug conjugates are often of disappointingly low toxicity.

Plant toxins, on the other hand, may be extremely toxic and remain so after conjugation, and there seems little risk of a cell becoming resistant to them. They act enzymatically, and it has been suggested that entry of a single molecule may be sufficient to kill a cell (Eiklid *et al.* 1980). They are, in general, easily conjugated with antibody, but if, like ricin, the toxin possesses a lectin B chain, the conjugate is likely to bind to many normal cells. This can be avoided either by preparing the conjugate in such a way that the lectin site is inactive, or by using a subunit of the toxin molecule from which the B chain has been cleaved enzymatically (Thorpe and Ross 1982). Immunotoxins lacking the B chain may not readily gain entry to cells, but various ways of overcoming this limitation have been proposed. Obviously, the more toxic the conjugate the more stringent the requirement for specificity of the antibody.

Conjugates of antibodies and radionuclides, including those that emit only alpha or beta particles, may be cytocidal even when they are outside the cell if they are within a critical distance, which depends on the radionuclide used and ranges from one to several hundred cell diameters (Humm 1986). If the radionuclide also emits gamma radiation the location of the conjugate may be determined with a gamma camera. A disadvantage of radionuclide conjugates is that before the conjugate becomes localized in the tumour the indiscriminate irradiation received by the patient may cause significant damage to bone marrow and intestinal epithelium.

Even if a conjugate fulfils the conditions discussed above, and can selectively kill tumour cells *in vitro*, there are two further factors that may prevent it from being therapeutically effective.

Firstly, it may not remain stable sufficiently long *in vivo*. Some conjugates are insufficiently stable *ab initio* to be useful. Even when this is not the case, if repeated doses are needed the patient may produce an antibody that reacts with the conjugate and causes it to be rapidly eliminated (see Levy and Miller 1983). One way of reducing this risk would be to use mAbs produced by human instead of rodent cells, but so far conjugates with human antitumour mAbs have been of low specificity. A compromise that appears promising is to use a human–rat hybrid antibody (Reichman *et al.* 1988). Even the use of human mAbs, however, does not dispose of the problem because the host antibody may be directed against the ideotype of the conjugated antibody, or the toxin moiety of the conjugate may behave as a hapten. Another approach, which Embleton (1987) suggests may prove to be the eventual choice, is to use Fab or $F(ab')_2$ fragments of rodent antibodies instead of the whole immunoglobulin molecule. Alternatively, one might try to protect the antibody molecule in some way, or administer an immunosuppressive agent.

Secondly, the tumour cells may be heterogeneous in respect of their capacity to bind a particular antibody. In consequence, the treatment may fall far short of the ideal, which is to destroy all tumour stem cells, and if only end state cells are killed will be completely ineffective. The range of neoplastic cells attacked may be increased by using two or more conjugates with different antibody specificities, but some of the tumour cells may not bind antibody at all, or not sufficiently strongly for the treatment to be effective. With radionuclide conjugates, especially conjugates with relatively energetic beta emitters, this problem may be less serious because cells at some distance from the one bearing the conjugate may receive a cytocidal dose of radiation, but, as we have seen, there is the danger of damage to normal tissues before the conjugate becomes localized in the tumour.

In view of these difficulties, Bagshawe (1985, 1987, 1989) has developed a new approach in which a conjugate of an antibody with an appropriate enzyme is used to activate a pro-drug. The conjugate is administered first and when it has become localized the pro-drug is administered. In consequence, active drug is released within the tumour and is taken up by cells in the vicinity of the cells bearing the conjugate. As Bagshawe has pointed out, the pro-drug should be a stable substance of low toxicity that is converted by the enzyme into a diffusible, highly toxic substance which has a short half-life *in vivo*. The enzyme should ideally be of non-mammalian origin, and lack a human analogue.

It is hoped that it will soon be possible to use recombinant DNA technology to make a molecule with an enzyme moiety and the antigen-binding site of an antibody that can be used therapeutically instead of a conjugate. The feasibility of doing this has been demonstrated by Neuberger *et al.* (1984), who successfully introduced Ig-gene DNA that had been manipulated *in vitro* into myeloma cells, and in this way were able to produce hapten-specific antibody in which the Fc part of the molecule was replaced by an active enzyme moiety.

It has been established that tumour cells in tissue culture can be destroyed, and the growth of human tumour xenografts can be inhibited, with targeted cytotoxic conjugates (see Moller 1982; Davies and Crumpton 1982; Baldwin and Byers 1985). Similar conjugates have been used in patients without causing serious complications, but it is difficult to assess their effect on the cancer because of other treatment which the patients received. Investigations are in progress (Bagshawe 1989) with enzyme conjugates targeted to carcino-embryonic antigen (CEA) on colon cancer cells or human chorionic gonadotrophin (HCG) on chorion carcinoma cells, using an alkylating agent as the 'war head'.

There seems little doubt that conjugates will play an increasingly important role in purging autologous bone marrow of neoplastic cells before returning it to the patient (Section 6.4.2, p. 85). Their future role as therapeutic agents directly administered to patients is more difficult to predict, but in the writer's view it is also likely to be very important.

If all the stem cells in all neoplasms of a particular kind possessed a common tumour-associated surface antigen, conjugates with a mAb that recognized this antigen would indeed be magic bullets, and almost certainly highly cost-effective. A conjugate with an antibody that selectively recognized all the stem cells of a neoplasm in a particular patient would be equally effective so far as that patient was concerned, but would have to be prepared specially and would inevitably be more expensive.

While these conditions may be *sufficient* to permit the development of conjugates that can be described as curative in an absolute sense (Section 6.1), they may not be *necessary*. A radionuclide conjugate, as we have seen, may destroy cells in the vicinity of the cell to which it becomes attached, and may therefore kill stem cells not recognized by the antibody. Moreover, stem cells that escape destruction by the conjugate may, conceivably, be destroyed by natural defence mechanisms. It is also important to remember that a combination of different agents may destroy all the stem cells of a tumour even though none of the agents acting alone can do this. Finally, there are the exciting, and as yet virtually unexplored, possibilities of using enzyme conjugates to activate pro-drugs in human patients, and of developing agents that combine the two essential properties of a conjugate in a single molecule.

7.4 Other immunological procedures

7.4.1 Introduction

In addition to the procedures discussed in the preceding section, attempts have been made to treat cancer by adoptive immunization with cells and cell products;

non-specific immunopotentiation with what are now called *biological response modifiers*, including BCG, *C. parvum*, synthetic polynucleotides, interferon and interleukin 2 (IL-2); and active immunotherapy with tumour cells or antigen. An account of this work up to 1980 will be found in Woodruff (1980), Chapters 7 and 8.

Interferon (rev. Goldstein and Laslo 1986) has generated great interest and, at times, great optimism. It has been consistently beneficial in hairy cell leukaemia, but overall it has proved disappointing; indeed it is open to question whether, up to now, the results are significantly better than those obtained with first generation biological response modifiers such as *C. parvum*, at much lower cost and with rather similar side-effects (see, for example, Woodruff and Walbaum 1983).

IL-2, originally called T-cell growth factor, is produced by T_H cells in the presence of macrophages. It too has so far proved disappointing as a therapeutic agent when used alone. Rosenberg *et al.* (1987) obtained one complete remission, which was still maintained after four months, and six partial remissions, in 46 patients with advanced cancer who received two five-day courses of IL-2 by intravenous injection every eight hours. The prescribed dose was 100 000 units per kg body weight per injection, but doses were sometimes reduced or omitted on account of toxic manifestations. These occurred in many patients and included fever, nausea and vomiting, hypotension, neutropenia and thrombocytopenia, and there were four deaths which were regarded as treatment related. Although these manifestations were partly controlled by appropriate treatment and usually disappeared soon after administration of IL-2 was discontinued, the severity of the side-effects in relation to the therapeutic effect is discouraging. Administration of IL-2 by continuous intravenous infusion has been reported to be as effective as intermittent injection but more comfortable for the patient (West *et al.* 1982).

More encouraging results have been obtained by adding adoptive immunization with lymphokine-activated killer cells (LAK cells), and it seems possible that tumour-infiltrating lymphocytes (TIL) will be even more effective. We shall discuss these procedures in the next section (7.4.2).

7.4.2 Adoptive immunization with LAK cells or TIL

Experimental background

Incubation of murine spleen cells (Rosenstein *et al.* 1984) or human peripheral blood lymphocytes (Lotze *et al.* 1981; Grimm *et al.* 1982) with IL-2 generates lymphokine-activated (LAK) cells that can lyse autologous or isologous tumour cells, including cells resistant to lysis by NK cells, *in vitro* while sparing normal cells.

Incubation of murine spleen cells with IL-4 (sometimes termed B-cell stimulatory factor 1) also generates LAK cells, and has been shown to augment the generation of LAK cytolytic activity induced by IL-2 (Mule et al. 1987).

Intravenous administration of LAK cells together with IL-2 has been shown to cause regression of a variety of immunogenic and non-immunogenic tumours in mice (Mule et al. 1984, 1985; Lafreniere and Rosenberg 1985a,b). Systemically-injected LAK cells can be 'focused' at the site of a tumour by local injection of products of macrophage activation such as interleukin 1 (IL-1) and tumour necrosis factor (Migliori et al. 1987).

Lymphocytes are commonly found in solid tumours (Section 1.2.1), and a population of TIL can be harvested and expanded by culture in the presence of IL-2 (Yron et al. 1980). Rosenberg et al. (1986) have investigated the use of such cells for adoptive immunotherapy. In mice with microscopic 'artificial metastases' produced by intravenous injection of tumour cells they were found to be 50–100 times more effective therapeutically than LAK cells. They also caused complete regression of large artificial metastases if the mice were given cyclophosphamide or whole-body irradiation (500 rad) simultaneously; these agents appear to act mainly by eliminating suppressor cells, although in the case of cyclophosphamide there may have been, in addition, some direct anti-tumour effect. Tumour regression was potentiated if the infusion of TIL was followed by injection of relatively small doses of IL-2, but could occur without this (Rosenberg et al. 1986).

TIL have also been isolated from human tumours, and tested for cytotoxic activity *in vitro*.

Itoh et al. (1986) found that about 4 per cent of TIL freshly harvested from human metastatic melanomas were able to bind to autologous tumour cells but did not appear to be cytotoxic. After two days in tissue culture the proportion of tumour-binding cells had increased and 10 per cent of cells were cytotoxic, and the proportions of tumour-binding and cytotoxic cells were increased further by continued culture in the presence of IL-2.

Muul et al. (1987) studied IL-2-stimulated TIL from six patients with melanoma. The cells from three patients were strongly cytotoxic for fresh autologous melanoma cells but were not toxic for normal peripheral blood lymphocytes or allogeneic tumour cells. Cells from the other patients were cytotoxic for autologous, and some allogeneic, melanoma cells, though not for normal cells.

Belldegrun et al. (1988) found that stimulated TIL from renal carcinomas were cytotoxic for almost all autologous tumour cells, and for cells from some allogeneic renal and non-renal tumours.

Clinical trials

In the light of the experimental results discussed in the preceding section, Rosenberg and his colleagues set up a Phase I study of LAK cells in patients with advanced cancer (Mazumder et al. 1984), followed by randomized trials in patients with metastatic cancer in whom standard treatment had failed or for whom no standard effective therapy was available (Rosenberg et al. 1985a,b, 1987). In both randomized trials objective remissions, and also toxic manifestations, were observed.

In the second and larger trial the patients received recombinant IL-2 by intravenous injection every eight hours for five days (100 000 units per kg body weight per dose), subject to the proviso that doses were reduced or omitted if this seemed desirable on account of toxic manifestations of the kind described in Section 7.4.1. After two days rest the patients were subjected to leucophoresis daily for five days. The mononuclear cells were separated on a density gradient, cultured *in vitro* in the presence of IL-2 for three or four days, and then returned to the patient, usually via a central venous catheter but in three patients into the hepatic artery via a catheter inserted percutaneously. Administration of IL-2 was continued 'until dose-limiting toxicity supervened'. Each leucophoresis lasted approximately four hours, during which time 10–14 litres of blood was processed. During the time when LAK cells were being infused the patients were kept in an intensive care unit, where continuous monitoring of cardiac function and central venous pressure was performed routinely, and various drugs were administered, including pressor and anti-inflammatory agents. These details are mentioned to give an idea of the magnitude of the procedure.

Of 106 evaluable patients, eight had a complete remission (median duration 10 months), during which all measurable tumour disappeared, and 25 had what were classified as partial (15) or minor (10) remissions. The best results were seen in patients with renal-cell carcinoma, melanoma, colorectal cancer and non-Hodgkin's lymphoma.

Rosenberg et al. (1987) conclude that adoptive immunotherapy with LAK cells plus IL-2, and to a lesser extent administration of IL-2 alone, 'can result in tumour regression [but without, of course, claiming that this is permanent] in some patients for whom no other effective treatment is available'. They have set up a further trial of adoptive immunotherapy in patients with smaller tumour burdens, but the results have not yet been reported.

Trials of lymphokine-activated TIL in combination with IL-2 and cyclophosphamide have been set up. In the light of the experimental results described in the preceding section there is good reason to expect that it will be possible to achieve a greater antitumour effect with smaller doses of IL-2, and therefore with less severe toxic manifestations, than have been obtained with LAK cells.

Whether these complicated procedures will contribute eventually to curing patients with cancer, as defined in Section 6.1, is an open question.

The answer may depend partly on whether ways can be found of reducing the toxic effects of IL-2 administration without weakening the therapeutic effect, and the use of TIL instead of LAK cells appears to offer some prospect of achieving this. Another possibility, discussed by Ezzell (1988), is to use recombinant DNA technology to 'engineer' whatever cells are used, before returning them to the patient, so that they produce their own IL-2 in sufficient quantity to obviate the need to administer IL-2 systemically to the patient, but it may prove difficult, and perhaps for the time being impossible, to obtain permission from the appropriate authorities to do this. As Ezzell points out, the only recorded experiments in which genetically-engineered cells have been used in the treatment of patients were those performed in the United States without such authorization by M. J. Cline in 1980; not surprisingly these caused a considerable furore and Dr Cline was made ineligible to receive federal government funds for research.

Another consideration, which is more fundamental and, in the writer's view, likely to prove a more serious obstacle, is whether neoplastic cells resistant to LAK cells or activated TIL exist in tumours *ab initio* or emerge in the course of treatment; and, if they do, whether such cells can be destroyed by other forms of treatment used in conjunction with adoptive immunotherapy.

7.5 Use of growth factors to combat damage to normal tissues caused by cytotoxic drugs

Reference has already been made (Section 2.6.1, pp. 25–26) to the work of Metcalfe and Sachs on the identification of haemopoietic growth factors, often called colony stimulating factors (CSF), and the possibility of using these to induce differentiation in myeloid leukaemic cells.

Another possible therapeutic role for these factors is to promote bone marrow recovery in patients with impaired marrow function caused by disease or the administration of cytotoxic drugs (rev. Thatcher 1989). Two factors, granulocyte CSF (G-CSF) and granulocyte-macrophage CSF (GM-CSF), both of which can be produced by recombinant DNA technology, are currently of particular interest.

Infusion of recombinant G-CSF has been reported to improve neutrophil function, and reduce the incidence of severe recurrent infection, in patients with small cell carcinoma of the lung receiving cytotoxic drugs, without causing toxic side-effects (Bronchud *et al.* 1987), and also to be beneficial in patients with advanced breast and ovarian cancer treated with doxorubicin in high dosage (Bronchud *et al.* 1989). Infusion of recombinant GM-CSF has been reported to lead to significant improvement in the leucocyte and platelet counts in patients with various metastatic tumours. Initially, this was associated with pyrexia, bone pain and pruritis, but in a later trial (Steward *et al.* 1989), in patients with metastatic colo-rectal cancer treated with melphalan in high dosage, a regimen of GM-CSF administration was found which shortened the duration of neutropenia and thrombocytopenia without causing toxic symptoms.

These and other investigations have established a *prima facie* case for the beneficial effect of administration of CSFs in marrow failure, but much remains to be learned concerning the optimal dosage, the degree of benefit that can be expected, and possible long-term complications.

It has been suggested that epidermal growth factor might be useful for the treatment of ulceration in the buccal cavity caused by cytotoxic drugs. This raises the further possibility that a growth factor might be found which was of value in patients with ulceration lower down in the alimentary tract.

7.6 Anti-angioneogenesis

The acquisition of a vascularized stroma is a critical step in the development of a primary solid tumour (Section 1.2.1) and also in the development of metastases (Section 5.2.2). Tumour angiogenesis factor (TAF) plays a key role in this process, and Folkman, who discovered TAF, suggested many years ago (Folkman 1972, 1974) that agents which inhibited the action of TAF might have a role in the treatment of cancer.

Evidence in support of this prediction has been provided by animal experiments in which an anti-TAF effect was achieved by administering cortisone in association with either heparin or a heparin fragment obtained by treating commercial pig heparin with bacterial heparinase (Folkman *et al.* 1983). It has since been shown that heparin binds preferentially to proliferating vascular endothelial cells, and this enhances the non-specific binding of cortisone to the cells (Sakamoto and Tanaka 1988).

In subsequent experiments Folkman and his colleagues (Crum *et al.* 1985) have shown that some corticosteroids devoid of glucocorticoid activity, notably a stereoisomer of hydrocortisone (epicortisol) and a tetrohydro derivative of cortisone (Tetrahydro S, Sigma), used in conjunction with heparin or heparin fragment, may be even more effective and less productive of unwanted sideeffects. This suggests that the time may be ripe for a cautious clinical trial of these agents in association with other forms of treatment of solid tumours (see Beranek 1988).

7.7 The problem of occult residual metastatic cancer

After complete ablation of a primary neoplasm in a patient who appears to be free of metastases, there are four possibilities to consider:

1. The patient is completely tumour-free.

2. A more thorough search with currently available techniques would reveal the presence of metastases.

3. There are undetectable metastases in which the neoplastic cell population is expanding quite rapidly, and which will soon become manifest.

4. Undetectable metastases are present but they are all dormant (Section 5.4).

The neoplastic cells are few in number and most of the stem cells are likely to be in G_0 or cycling very slowly.

In the last two cases the patient may be said to have occult residual metastatic cancer. This situation can only be diagnosed retrospectively, if and when one or more metastases become manifest, though in the light of experience of patients with similar types of cancer it may be possible to estimate the likelihood that metastases will appear sooner or later.

The threat of occult residual metastases prompted the introduction of systemic prophylactic therapy, generally referred to as adjuvant therapy (Section 6.4.2, pp. 82–84). In an individual patient the appearance of metastases shows that this treatment has failed, but there is no way of knowing if it has been successful. Statistical evidence, obtained by comparing the fate of patients with similar tumours who did, or did not, receive adjuvant therapy, shows that it reduces the incidence of metastases in some categories of patient, notably premenopausal women with breast cancer; on the whole, however, as we have seen, the results have been disappointing, and the risk of inducing new neoplasms in young patients cannot be entirely dismissed.

One reason for the optimism engendered by the introduction of adjuvant therapy is that if occult metastases do exist the tumour burden is small, but against this must be set the disadvantage that there is no way of monitoring the effect of the treatment until it has failed, and it may be doing harm by causing drug-resistant cells to develop, or even by promoting the appearance of metastases that would otherwise have remained dormant indefinitely. This has led some clinicians to give systemic therapy before local treatment (Forrest *et al*. 1986)—a procedure sometimes called neo-adjuvant therapy. Monitoring is now possible, but regression of the primary tumour, though encouraging, does not necessarily mean that the cells in occult metastases, which may be phenotypically very different from the majority of cells in the primary tumour, will be similarly affected. Moreover, the advantage of a small tumour burden, if it really is an advantage, is lost.

If adjuvant therapy is to be given, what form should it take?

Cells that are in G_0, or are cycling very slowly, are likely to be refractory to many antimetabolites and mitotic inhibitors, and also to DNA binding agents if the cells can repair DNA lesions before DNA synthesis and cell division are resumed (Skipper and Schabel 1984). The use of other types of cytotoxic agent, and of multiple agents, should sharpen the attack, but it is difficult to balance the side-effects that can be seen in individual patients with possible benefits that can be estimated only on a statistical basis.

Another approach that might be considered is to try to precipitate the appearance of metastases as a preliminary to giving cytotoxic agents when the tumour bulk, though larger than it was before the metastases appeared, is still fairly small, but the risks would seem to be large in relation to the possible benefit.

In the writer's view, there is a strong case for undertaking clinical trials of adjuvant therapy with some of the new agents discussed earlier in this chapter, but it may be difficult to recruit enough patients and doctors. Informed patient consent to participate in any trial of adjuvant therapy implies that the patient has been told that there is a serious possibility, but no certainty, that residual disease exists; and has been given a realistic assessment of the prognosis without adjuvant therapy, and an honest opinion concerning the possible benefits and the disadvantages of such therapy. Some doctors may feel that it is not in the interest of their patient to embark on a discussion of this kind, at least until the position has become clearer.

Epilogue

The dramatic advances that have occurred in biology in the last two decades have generated, among other things, a vast amount of information on which to base new applied research concerning the diagnosis and treatment of cancer. It might seem to some that, for the time being, available resources in trained researchers and materials should be concentrated on this task and further basic research should be given a lower priority. To adopt such a policy would, however, be disastrous. Many, including the writer (Woodruff 1977, 1987), have argued the general case for the importance of basic research. The *ad hoc* argument that progress in basic research should be slowed down to give us time to catch up on the applied front is fallacious because applied research needs to be based on the fullest possible information, and momentum, once lost, would be difficult to regain.

This thesis is well illustrated by the subject of the present book. The extent to which cancer cell heterogeneity and adaptation limit the effectiveness of current methods of treating cancer is becoming increasingly clear, and there is, indeed, much to be done at the level of applied research in developing new therapeutic tactics. But to achieve a major strategical advance we will need a still deeper understanding of the biological basis of these phenomena. The history of science leaves no room for doubt that knowledge of this kind comes from the work of people driven by insatiable curiosity to search for truth for its own sake, and who know that there will always be more truth to be discovered.

References

Abercrombie, M. (1980). The Croonian Lecture, 1977. The crawling movement of metazoan cells. *Proc. R. Soc. Lond.* B **207**, 129–47.
Abercrombie, M. and Heaysman, J. E. M. (1954). Observations on the social behaviour of cells in tissue culture. *Expl. Cell Res.* **6**, 293–306.
Adams, G. E., Dische, S., Fowler, J. F., and Thomlinson, R. H. (1976). Hypoxic cell sensitisers in radiotherapy. *Lancet* **1**, 186–8.
Adams, J. M., Harris, A. W., Pinkert, C. A., Cochran, L. M., Alexander, W. S., Cory, S., Palmiter, R. D., and Brinster, R. L. (1985). The c-*myc* oncogene driven by immunoglobulin enhancers induces lymphoid malignancy in transgenic mice. *Nature* **318**, 533–8.
Ahern, W. A., Camplejohn, R. S., and Wright, N. A. (1977). *An introduction to cell population kinetics*. Edward Arnold, London.
Albino, A. P., Le Strange, R., Oliff, A. I., Furth, M. E., and Old, L. J. (1984). Transforming *ras* genes from human melanoma: a manifestation of tumour heterogeneity? *Nature* **308**, 69–72.
Alexander, P. (1982). Control of metastatic spread by the immune defences of the host. *Clinics in Oncology* **1**, 620–35.
Alexander, P. (1985). Commentary. Do cancers arise from a single transformed cell or is monoclonality of tumours a late event in carcinogenesis? *Br. J. Cancer* **51**, 453–7.
Alitalo, K. and Schwab, M. (1986). Oncogene amplification in tumor cells. *Adv. Cancer Res.* **47**, 235–81.
Altevogt, P., von Hoegen, P., and Schirrmacher, V. (1986). Immunoresistant metastatic tumor variants can re-express their tumor antigen after treatment with DNA methylation-inhibiting agents. *Int. J. Cancer* **38**, 707–11.
Ansell, J. D., Hodson, B. A., and Woodruf, M. F. A. (1986). The provenance of cells in sarcomas induced in chimaeric mice. *Brit. J. Cancer* **54**, 853–5.
Ash, D. (1986). Interstitial radiation therapy—the current position. *Cancer Topics* **5**, 142.
Asseline, W., Delarue, M., Lancelot, G., Toulme, F., Thuong, N. T., Montenay-Garestier, T., and Helene, C. (1984). *Proc. Natl. Acad. Sci. USA* **81**, 3297–301.
Bagshawe, K. D. (1985). Cancer drug targeting. *Clin. Radiol.* **36**, 545–51.
Bagshawe, K. D. (1987). Antibody directed enzymes revive anti-cancer prodrugs concept. *Br. J. Cancer* **56**, 531–2.
Bagshawe, K. D. (1989). The Bagshawe Lecture. While waiting for the human genome to be mapped. *Brit. J. Cancer* **60**, 275–81.

Baldwin, R. W. and Byers, V. S. (ed.) (1985). *Monoclonal antibodies for cancer detection and therapy*. Academic Press, London.

Balmain, A. (1985). Transforming *ras* oncogenes and multistage carcinogenesis. *Brit. J. Cancer* **51**, 1-7.

Baltimore, D. (1981). Somatic mutation gains its place among the generators of diversity. *Cell* **26**, 295-6.

Barker, D., Wright, E., Nguyen, K., Cannon, L., Fain, P., Goldgar, D., Bishop, D. T., Carey, B., Baty, B., Kivlin, J., Willard, H., Waye, J. S., Greig, G., Leinwand, L., Nakamura, Y., O'Connell, P., Leppert, M., Lalouel, J-M., White, R., and Skolnick, M. (1987). Gene for von Recklinghausen Neurofibromatosis is in the pericentromeric region of chromosome 17. *Science* **236**, 1100-2.

Bassin, R. H. and Noda, M. (1987). Oncogene inhibition by cellular genes. *Adv. Viral. Oncol.* **6**, 103-27.

Batterman, J. J. and Mijnheer, B. J. (1986). The Amsterdam fast neutron therapy project. *Int. J. Radiat. Oncol. Biol. Phys.* **12**, 2093-9.

Belldegrun, A., Muul, L. M., and Rosenberg, S. A. (1988). Interleukin 2 expanded tumor-infiltrating lymphocytes in human renal cell cancer: isolation, characterization, and antitumor activity. *Cancer Res.* **48**, 206-14.

Beranek, J. T. (1988). Antiangiogenesis comes out of its shell. *The Cancer Journal* **2**, 87-8.

Berenbaum, M. C. (1968). The last surviving cancer cell: the chances of killing it. *Cancer Chemoth. Rep.* **52**, 539-41.

Berenbaum, M. C. (1969). Dose-response curves for agents that impair cell reproductive integrity. *Br. J. Cancer* **23**, 426-33: 434-45.

Berger, N. A., Berger, S. J., and Gerson, S. L. (1987). DNA repair. ADP-ribosylation and pyridine nucleotide metabolism as targets for cancer chemotherapy. *Anti-Cancer Drug Design* **2**, 203-9.

Bernhard, H. P. (1976). The control of gene expression in somatic cell hybrids. *Int. Rev. Cytol.* **47**, 289-325.

Bernstein, S. C. and Weinberg, R. A. (1985). Expression of the metastatic phenotype in cells transfected with human metastatic tumor DNA. *Proc. Natl. Acad. Sci. USA* **82**, 1726-30.

Bishop, J. M. (1981). Enemies within: the genesis of retrovirus oncogenes. *Cell* **23**, 5-6.

Bishop, J. M. (1987). The molecular genetics of cancer. *Science* **235**, 305-11.

Bissell, M. J. (1988). Extracellular matrix influence on gene expression: is structure the message? *Br. J. Cancer* **58**, 223. (abst.).

Bissell, M. J., Hall, H. G., and Parry, G. (1982). How does the extracellular matrix direct gene expression? *J. Theoret. Biol.* **99**, 31-68.

Blake, K. R., Murakami, A., Spitz, S. A., Glave, M. P., Reddy, M. P., Ts'o, P. O. P., and Miller, P. S. (1985). Hybridization arrest of globin synthesis in rabbit reticulocyte lysates and cells by oligodeoxyribonucleoside methylsulphonates. *Biochemistry* **24**, 6139-45.

Bobrow, M. (1988). The prevention and avoidance of genetic disease: summing up. *Phil. Trans. R. Soc. Lond. B* **319**, 361-7.

Bodenham, D. C. (1968). A study of 650 observed malignant melanomas in the South-West region. *Ann. R. Coll. Surg. Engl.* **43**, 218-39.
Bodmer, W. F., Bailey, C. J., Bodmer, J., Bussey, H. J. R., Ellis, A., Gorman, P., Lucibello, F. C., Murday, V. A., Rider, S. H., Scambler, P., Sheer, D., Solomon, E., and Spurr, N. K. (1987). Location of the gene for familial adenomatous polyposis on chromosome 5. *Nature* **328**, 614-16.
Bonadonna, G., Valagussa, B., Tancini, G., Rossi, A., Brambilla, C., Zambetti, M., Bignassi, P., Di Fronzo, G., and Silvestrini, R. (1986). Current status of Milan adjuvant chemotherapy trials of node-positive and node-negative breast cancer. *Natl. Cancer Inst. Monographs* **1**, 45-9.
Boon, T. (1983). Antigenic tumour cell variants obtained with mitogens. *Adv. Cancer Res.* **39**, 121-51.
Bos, J. L., Fearon, E. R., Hamilton, S. R., Verlaan-de Vries, M., van Boom, J. H., van de Eb, A. J., and Vogelstein, B. (1987). Prevalence of *ras* gene mutations in human colorectal cancers. *Nature* **327**, 293-7.
Bradley, G., Juranka, P. F., and Ling, V. (1988). Mechanism of multidrug resistance. *Biochim. Biophys. Acta* **948**, 87-128.
Broder, S. (1984). T-cell lymphoproliferative syndrome associated with human T-cell leukaemia/lymphoma virus. *Ann. Intern. Med.* **100**, 543-57.
Bronchud, M. H., Scarffe, J. H., Thatcher, N., Crowther, D., Souza, L. M., Alton, N. K., Testa, N. G., and Dexter, T. M. (1987). Phase I/II study of recombinant human granylocyte colony-stimulating factor in patients receiving intensive chemotherapy for small cell lung cancer. *Br. J. Cancer* **56**, 809-13.
Bronchud, M. H., Howell, A., Crowther, D., Hopwood, P., Souza, L., and Dexter, T. M. (1989). Two weekly high-dose doxorubicin therapy with infusions of granulocyte colony-stimulating factor in patients with advanced breast and ovarian cancer (abst.). *Br. J. Cancer* **60**, 449.
Burchenal, J. H. (1976). Adjuvant therapy—theory, practice and potential. *Cancer* **37**, 46-57.
Cairns, J. (1975). Mutation selection and the natural history of cancer. *Nature* **255**, 197-200.
Carr, B. I., Gilchrist, K. W., and Cavanee, P. (1981). The variable transformation in metastases from testicular germ cell tumours: the need for selective biopsy. *J. Urol.* **126**, 52-4.
Catterall, M., Bewley, D. K., and Sutherland, I. (1977). Second report on results of a randomised clinical trial of fast neutrons compared with X or gamma rays in treatment of advanced tumours of head and neck. *Br. Med. J.* **1**, 1642.
Cavanee, W. K., Dryja, T. P., Phillips, R. A., Benedict, W. F., Godbout, R., Gallie, B. L., Murphree, A. L., Strong, L. C., and White, R. L. (1983). Expression of recessive alleles by chromosomal mechanisms in retinoblastoma. *Nature* **305**, 779-84.
Chiu, C. P. and Blau, H. M. (1984). Programming cell differentiation in the absence of DNA synthesis. *Cell* **37**, 879-87.
Chow, D. A. and Greenberg, A. H. (1980). The generation of tumour heterogeneity *in vivo. Int. J. Cancer* **25**, 261-5.
Cole, W. H. (1973). The mechanisms of spread of cancer. *Surg. Gynecol. Obstet.* **137**, 853-71.

Coleman, M. P., Bell, C. M. J., and Fraser, P. (1987). Second primary malignancy after Hodgkin's disease, ovarian cancer and cancer of the testis: a population based cohort study. *Br. J. Cancer* **56**, 349–55.

Comings, D. E. (1973). A general theory of carcinogenesis. *Proc. Natl. Acad. Sci. USA* **70**, 3324–8.

Coombes, R. C., Powles, T. J., Berger, U., Wilson, P., McClelland, R. A., Gazet, J-C., Troitt, P. A., and Ford, H. T. (1987). Prediction of endocrine response in breast cancer by immunocytochemical detection of oestrogen receptor in fine-needle aspirates. *Lancet* **2**, 701–3.

Craig, R. W. and Sager, R. (1985). Suppression of tumorigenicity in hybrids of normal and oncogene-transferred CHEF cells. *Proc. Natl. Acad. Sci. USA* **82**, 2062–6.

Crum, R., Szabo, S., and Folkman, J. (1985). A new class of steroids inhibits angiogenesis in the presence of heparin or heparin fragment. *Science* **230**, 1375–8.

Cushing, H. and Wolback, S. B. (1927). The transformation of a malignant paravertebral sympathicoblastoma into a benign ganglioneuroma. *Am. J. Path.* **5**, 203–16.

Cuzick, J., Stewart, H., Peto, R., Fisher, B., Kase, S., Johansen, H., Lythgoe, J. P., and Prescott, R. J. (1987*a*). Overview of randomized trials comparing radical mastectomy without radiotherapy against simple mastectomy with radiotherapy in breast cancer. *Cancer Treat. Rep.* **71**, 7–13.

Cuzick, J., Stewart, H., Peto, R., Baum, M., Fisher, B., Host, H., Lythgoe, P., Ribeiro, G., Scheurlen, H., and Wallgren, A. (1987*b*). Overview of randomized trials of postoperative adjuvant radiotherapy in breast cancer. *Cancer Treat. Rep.* **71**, 15–29.

Daar, A. S. (ed.) (1987). *Tumour markers in clinical practice.* Blackwell Scientific Publications, Oxford.

Davies, A. J. S. and Crumpton, M. J. (ed.) (1982). Experimental approaches to drug targeting. *Cancer Surveys* **1**, 347–559.

Davis, N. C., McLeod, G. R., Beardmore, G. L., Little, J. H., Quinn, R. L., and Holt, J. (1976). Primary cutaneous melanoma: a report from the Queensland melanoma project. *CA Cancer J. for Clinicians* **26**, 81–107.

Deamant, F. D., Vidj, M., Rossant, J., and Iannoccone, P. M. (1986). *In situ* identification of host-derived infiltrating cells in chemically induced fibrosarcomas of interspecific chimeric mice. *Int. J. Cancer* **37**, 283–6.

De la Monte, S. M., Moore, G. W., and Hutching, G. M. (1983). Pattern distribution of metastases from malignant melanoma in humans. *Cancer Res.* **43**, 3427–34.

De Vita, V. T. (1983). The James Ewing Lecture. The relationship between tumor mass and resistance to chemotherapy: implications for surgical adjuvant treatment of cancer. *Cancer* **51**, 1209–20.

Dexter, D. L. and Calabresi, P. (1982). Intraneoplastic diversity. *Biochim. Biophys. Acta* **695**, 97–112.

Doreen, M. S., Wrigley, P. F. M., Laidlow, J. M., Plowman, P. M., Neudachin, L., Tucker, A. K., Malpas, J. S., Stansfield, A. G., Faux, M. M. L., Jones, A. E., and Lister, T. A. (1984). The management of Stage II supradiaphragmatic Hodgkin's disease at St. Bartholomew's Hospital. *Cancer* **54**, 2882–8.

Downward, J., Yarden, Y., Mayer, E., Scrace, G., Totty, N., Stockwell, P., Ullrich, A., Schlessinger, J., and Waterfield, M. D. (1984). Close similarity of epidermal growth factor receptor and v-*erb*-B oncogene protein sequences. *Nature* **307**, 521–7.

Dryja, T. P., Rapaport, J. M., Epstein, J., Goorin, A. M., Weichselbaum, R., Koufos, A. and Cavanee, W. K. (1986). Chromosome 13 homozygosity in osteosarcoma without retinoblastoma. *Am. J. Hum. Genet.* **38**, 59–66.

Duesberg, P. H. (1985). Activated proto-onc genes: Sufficient or necessary for cancer? *Sicence* **228**, 669–76.

Duesberg, P. H. (1987). Cancer genes: rare recombinants instead of activated oncogenes (a review). *Proc. Natl. Acad. Sci. USA* **84**, 2117–24.

Dulbecco, R. (1982). The nature of cancer. *Endeavour. New Series* **6**, 59–65.

Duncan, W., Arnott, S. J., Jack, W. J. L., Macdougall, R. H., Quilty, P. M., Rodger, A., Kerr, G. R., and Williams, J. R. (1985). A report on a randomized trial of d(15) + Be neutrons compared with megavoltage X-ray therapy of bladder cancer. *Int. J. Radiat. Oncol. Biol. Phys.* **11**, 2043–59.

Ebbs, S. R. (1987). Practical aspects of prolonged intravenous cytotoxic infusion therapy. *Cancer Topics* **6**, 106–7.

Eccles, S. A. and Alexander, P. (1975). Immunologically-mediated restraint of latent tumour metastases. *Nature* **257**, 52–3.

Edwards, P. A. W. (1985). Heterogeneous expression of cell-surface antigens in normal epithelia and their tumours, revealed by monoclonal antibodies. *Br. J. Cancer* **51**, 149–60.

Eiklid, K., Olsnes, S., and Pihl, A. (1980). Entry of lethal doses of abrin, ricin and modecein into the cytosol of HeLa cells. *Exp. Cell Res.* **126**, 321–6.

Ellman, R. (1986). Cervical cancer screening. *Cancer Topics* **5**, 122–3.

Ellman, R. (1987). Breast cancer screening. *J. R. Soc. Med.* **80**, 665–6.

Embleton, M. J. (1987). Editorial. Drug-targeting by monoclonal antibodies. *Br. J. Cancer* **55**, 227–31.

EORTC Breast Cancer Cooperative Group. (1980). Revision of standards for the assessment of hormone receptors in human breast cancer. Report of the second EORTC workshop. *Eur. J. Cancer Clin. Oncol.* **16**, 1513–15.

Evans, H. J. (1986). The role of human cytogenesis in studies of mutagenesis and carcinogenesis. In *Genetic toxicology of environmental chemicals. Part A: Basic principles and mechanisms of action* (ed. C. Ramel, B. Lambert, and J. Magnussen), pp. 41–69. Alan R. Liss Inc., New York.

Everson, T. C. and Cole, W. H. (1966). *Spontaneous regression of cancer.* W. B. Saunders, Philadelphia.

Ezzell, C. (1988). News. Plans for altered lymphocyte release in humans. *Nature* **333**, 697.

Farber, E. and Cameron, R. (1980). The sequential analysis of cancer development. *Adv. Cancer Res.* **31**, 125–226.

Feldman, M. and Eisenbach, L. (1988). What makes a tumor cell metastatic? *Scientific American* **259**, 40–7.

Fialkow, P. J. (1972). Use of genetic markers to study cellular origin and development of tumors in human females. *Adv. Cancer Res.* **15**, 191–225.

Fialkow, P. J. (1976). Clonal origin of human tumors. *Biochim. Biophys. Acta* **458**, 283–321.

Fidler, I. J. (1973). Selection of successive tumour lines for metastasis. *Nature* **242**, 148–9.

Fidler, I. J. and Hart, I. R. (1982). Biological diversity in metastatic neoplasms: origins and implications. *Science* **217**, 998–1003.

Fidler, I. J. and Kripke, M. L. (1977). Metastasis results from pre-existent variant cells within a malignant tumor. *Science* **198**, 893–5.

Fischinger, P. J. and Haapala, D. K. (1974). Oncoduction. A unifying hypothesis of viral carcinogenesis. *Prog. Exptl. Tumor Res.* **19**, 1–22.

Fisher, B. (1971). Status of adjuvant therapy: results of the National Surgical Adjuvant Breast Cancer Project studies on oophorectomy, post-operative radiation therapy, and chemotherapy. *Cancer* **28**, 1654–8.

Fisher, B. and Fisher, E. R. (1959). Experimental evidence in support of the dormant tumor cell. *Science* **130**, 918–19.

Fisher, B. and Fisher, E. R. (1966a). Transmigration of lymph nodes by tumor cells. *Science* **152**, 1397–8.

Fisher, B. and Fisher, E. R. (1966b). The interrelationship of hematogenous and lymphatic tumor dissemination. *Surg. Gynecol. Obstet.* **122**, 791–8.

Fisher, B., Rockette, H., Fisher, E. R., Wickerham, D. L., Redmond, C., and Brown, A. (1985). Leukemia in breast cancer patients following adjuvant chemotherapy or postoperative irradiation: the NSABP experience. *J. Clin. Oncol.* **3**, 1640–58.

Fisher, B., Fisher, E. R., and Redmond, C. (1986). Ten-year results from the National Surgical Adjuvant Breast and Bowel Project (NSABP) clinical trial evaluating the use of L-phenylalanine mustard (L-PAM) in the management of primary breast cancer. *J. Clin. Oncol.* **4**, 929–41.

Fiszman, M. Y. and Fuchs, P. (1975). Temperature-sensitive expression of differentiation in transformed myoblasts. *Nature* **254**, 429–31.

Folkman, J. (1972). Antiangiogenesis: new concept for therapy of solid tumors. *Ann. Surg.* **175**, 409–16.

Folkman, J. (1974). Tumor angiogenesis. *Adv. Cancer Res.* **19**, 331–58.

Folkman, J., Langer, R., Linhardt, R. J., Haudenschild, C., and Taylor, S. (1983). Angiogenesis inhibition and tumor regression caused by heparin or heparin fragment in the presence of cortisone. *Science* **221**, 719–25.

Forrest, A. P. M., Levack, P. A., Chetty, U., Hawkins, R. A., Miller, W. R., Smyth, J. F., and Anderson, T. J. (1986). A human tumour model. *Lancet* **2**, 840–1.

Forrester, K., Almoguera, C., Han, K., Grizzle, W. E., and Perucho, M. (1987). Detection of high incidence of K-*ras* oncogenes during human colon tumorigenesis. *Nature* **327**, 298–303.

Foulds, L. (1969). *Neoplastic development*. Academic Press, London.

Fournier, R. E. K. and Ruddle, F. H. (1977). Microcell-mediated transfer of murine chromosomes into mouse, Chinese hamster, and human somatic cells. *Proc. Natl. Acad. Sci. USA* **74**, 319–23.

Fox, K. R. and Waring, M. J. (1984). DNA structural variations produced by actinomycin and distamycin as revealed by DNA footprinting. *Nucleic Acids Res.* **12**, 9271–85.

Fox, K. R., Davies, H., Gill, R. A., Portugal, J. and Waring, M. J. (1988). Sequence-specific binding of luzpeptin to DNA. (1988). *Nucleic Acids Res.* **6**, 2489–507.

Franks, L. M. and Teich, N. (1986). *Introduction to the cellular and molecular biology of cancer*. Oxford University Press, Oxford.

Friend, S. H., Bernards, R., Rogelj, S., Weinberg, R. A., Rapaport, J. M., Albert,

D. M., and Dryja, T. P. (1986). A human DNA segment with properties of the gene that predisposes to retinoblastoma and osteosarcoma. *Nature* **323**, 643–6.

Fristrom, J. W. and Spieth, P. T. (1980). *Principles of genetics.* Chiron Press, New York; and Blackwell Scientific Publications, Oxford.

Garnett, M. C. and Baldwin, R. W. (1986). An improved synthesis of a methotrexate-albumin-791T/36 monoclonal antibody conjugate cytotoxic to human osteogenic sarcoma cell lines. *Cancer Res.* **46**, 2407–12.

Geiser, A. G. Der, C. J., Marshall, C. J., and Stanbridge, E. J. (1986). Suppression of tumorigenicity with continued expression of the c-Ha-*ras* oncogenes in EJ bladder carcinoma–human fibroblast hybrid cells. *Proc. Natl. Acad. Sci. USA* **83**, 5209–13.

Giavazzi, R., Allesandri, G., Spreafico, F., Garattini, S., and Mantovani, A. (1980). Metastasizing capacity of tumour cells from spontaneous metastases of transplanted murine tumours. *Br. J. Cancer* **42**, 462–72.

Globerson, A. and Feldman, M. (1964). Antigenic specificity of benzo(a)pyrene-induced sarcomas. *J. Natl. Cancer Inst.* **32**, 1229–43.

Goldie, J. H. and Coldman, A. J. (1984). The genetic origin of drug resistance on neoplasms: implications for systemic therapy. *Cancer Res.* **44**, 3643–53.

Goldstein, D. and Laszlo, J. (1986). Interferon therapy in cancer: from imaginon to interferon. *Cancer Res.* **46**, 4315–29.

Gootwine, E., Webb, C. G., and Sachs, L. (1982). Participation of myeloid leukemic cells injected into embryos in haematopoietic differentiation in adult mice. *Nature* **299**, 63–5.

Gottesman, M. M. and Pastan, I. (1988). The multidrug transporter, a double edged sword. *J. Biol. Chem.* **263**, 12163–6.

Goyns, M. H. (1986). The role of oncogenes in human cancer. *The Cancer Journal* **1**, 183–90.

Gray, B. N. (1984). Treatment of liver metastases from large bowel cancer. In *Integrated management of cancer* (ed. R. M. Nambiar and G. J. A. Clunie), pp. 115–29. Singapore Cancer Society, Singapore.

Green, H. S. N. (1950). Personal communication.

Greig, R. G., Koestler, T. P., Trainer, D. L., Corwin, S. P., Miles, L., Kline, T., Sweet, R., Yokoyama, S., and Poste, G. (1985). Tumorigenic and metastatic properties of 'normal' and *ras*-transfected NIH/3T3 cells. (1985). *Proc. Natl. Acad. Sci. USA* **82**, 3698–701.

Grimm, E. A., Mazumder, A., Zhang, H. Z., and Rosenberg, S. A. (1982). Lymphokine-activated killer cell phenomenon: lysis of natural-killer resistant fresh solid tumor cells by interleukin 2-activated autologous human peripheral blood lymphocytes. *J. Exper. Med.* **155**, 1823–41.

Groner, B., Schonenberger, C-A., and Andres, A. C. (1987). Targeted expression of the *ras* and *myc* oncogenes in transgenic mice. *Trends in Genet.* **3**, 306–8.

Gullick, W. and Sikora, K. (1988). Oncogenes as clinical tools. *Cancer Topics* **6**, 138–40.

Gurdon, J. B. (1977). The Croonian Lecture, 1976. Egg cytoplasm and gene control in development. *Proc. R. Soc. Lond., B.* **198**, 211–47.

Halliday, R., Porterfield, J. S., and Gibbs, D. D. (1974). Premature ageing and occurrence of altered enzyme in Werner's syndrome fibroblasts. *Nature* **248**, 762–3.

Handley, W. S. (1922). *Cancer of the breast.* John Murray, London.

Harrap, K. R. and Connors, T. A. (eds) (1987). *New avenues in developmental cancer chemotherapy*. Part I. Academic Press, London.

Harris, A. L. (1988). DNA repair and drug resistance. *Cancer Tropics* **6**, 111-12.

Harris, H. (1971). Cell fusion and the analysis of malignancy. *Proc. R. Soc. Lond. B.* **179**, 1-20.

Harris, H. (1985). The suppression of malignancy in hybrid cells: The mechanism. *J. Cell Sci.* **79**, 83-94.

Harris, J. F., Chambers, A. F., Hill, R. P., and Ling, V. (1982). Metastatic variants are generated spontaneously at a high rate in mouse KHT tumor. *Proc. Natl. Acad. Sci. USA* **79**, 5547-51.

Hawkins, M. M., Draper, D. J., and Kingston, J. E. (1987). Incidence of second primary tumours among childhood cancer survivors. *Br. J. Cancer* **56**, 339-47.

Haybittle, J. L. (1964). The cured group in series of patients treated for cancer. *Anglo-German Med. Rev.* **2**, 422-36.

Haybittle, J. L., Hayhoe, F. G. J., Easterling, M. J., Jeliffe, A. M., Bennett, M. H., Hudson, G. V., Hudson, R. V., and MacLennan, K. A. (1985). Review of British National Lymphoma Investigation studies of Hodgkin's disease and development of a prognostic index. *Lancet* **1**, 967-72.

Hayward, J. L. and Rubens, R. D. (1987). UICC Multidisciplinary project on breast cancer. Management of early and advanced breast cancer. *Int. J. Cancer* **39**, 1-5.

Helene, C. (1988). The 1988 Walter Hubert Lecture. Artificial regulation of gene expression by oligonucleotides covalently linked to intercalating agents. *Br. J. Cancer* **58**, 225. (abst.).

Hellmann, K., Gilbert, J., Evans, M., Cassell, P., and Taylor, R. (1987). Effect of razoxane on metastases from colorectal cancer. *Clin. Expl. Metastasis* **5**, 3-8.

Henk, J. M. and Smith, C. W. (1977). Radiotherapy and hyperbaric oxygen in head and neck cancer. Interim report of a second clinical trial. *Lancet* **2**, 104-5.

Heppner, G. H. (1982). Tumor subpopulation interactions. In *Tumor cell heterogeneity* (ed. A. H. Owens, D. S. Coffey and S. B. Baylin), pp. 225-35.

Heppner, G. H. (1984). Tumor heterogeneity. *Cancer Res.* **44**, 2259-65.

Hermans, A., Heisterkamp, N., von Lindern, M., van Baal, S., Meijer, D., van der Plas, D., Wiedemann, L. M., Groffen, J., Bootsma, D., and Grosveld, G. (1987). Unique fusion of *bcr* and *c-abl* genes in Philadelphia chromosome positive acute lymphoblastic leukemia. *Cell* **51**, 33-40.

Hethcote, H. W. and Knudson, A. G. (1978). Model for the incidence of embryonal cancers: application to retinoblastoma. *Proc. Natl. Acad. Sci. USA* **75**, 2453-7.

Hewitt, H. B. and Blake, E. (1975). Quantitative studies of translymphnodal passage of tumour cells naturally disseminated from a non-immunogenic murine squamous carcinoma. *Br. J. Cancer: 312*, 25-35.

Hill, M. and Hillova, J. (1976). Genetic transformation of animal cells with viral DNA or RNA tumor viruses. *Adv. Cancer Res.* **23**, 237-97.

Hill, R. P., Chambers, A. F., Ling, V., and Harris, J. F. (1984). Dynamic heterogeneity: Rapid generation of metastatic variants in mouse B16 melanoma cells. *Science* **224**, 998-1001.

Howell, M. B., Wareham, K. A., and Williams, E. D. (1985). Clonal origin of mouse liver tumors. *Amer. J. Pathol.* **121**, 426-32.

Hsu, S. H., Luk, G. D., Krush, A. J., Hamilton, S. R., and Hoover, H. H. (1983). Multiclonal origin of polyps in Gardner syndrome. *Science* **221**, 951-3.
Huebner, R. J. and Todaro, G. J. (1969). Oncogenes of RNA tumor viruses as determinants of cancer. *Proc. Natl. Acad. Sci. USA* **64**, 1087-94.
Huggins, C. (1967). Endocrine-induced regression of cancers. *Science* **156**, 1050-4.
Humm, J. L. (1986). Dosimetric aspects of radiolabeled antibodies for tumor therapy. *J. Nucl. Med.* **27**, 1490-1497.
Hurwitz, E., Maron, R., Bernstein, A., Wilchek, M., Sela, M., and Arnon, R. (1978). The effect *in vivo* of chemotherapeutic drug-antibody conjugates in two murine experimental tumor systems. *Int. J. Cancer* **21**, 747-55.
Iannaccone, P. M., Weinberg, W. C. and Deamant, F. D. (1987). On the clonal origin of tumors: a review of experimental models. *Int. J. Cancer* **39**, 778-84.
Illmensee, K. and Hoppe, E. (1981). Nuclear transplantation in *Mus musculus*: developmental potential of nuclei from pre-implantation embryos. *Cell* **23**, 9-18.
Illmensee, K. and Mintz, B. (1976). Totipotency and normal differentiation of single teratocarcinoma cells cloned by injection in blastocysts. *Proc. Natl. Acad. Sci. USA* **73**, 549-53.
Itoh, K., Tilden, A. B., and Balch, C. M. (1986). Interleukin 2 activation of cytotoxic T-lymphocytes infiltrating into human metastatic melanomas. *Cancer Res.* **46**, 3011-17.
Johnsson, A., Heldin, C. H., Wasteson, A., Westermark, B., Devel, T. F., Huang, T. S., Seeburg, P. H., Gray, A., Ullrich, A., Scrace, G., Stroobant, P., and Waterfield, M. D. (1984). The c-*cis* gene encodes a precursor of the B chain of PDGF. *EMBO J.* **3**, 921-3.
Jonat, W., Maass, H., and Stegner, H. E. (1986). Immunohistochemical measurement assay of estrogen receptors in breast cancer tissue samples. *Cancer Res.* (Suppl.) **46**, 4296-8.
Kaldor, J. M., Day, N. E., Band, P., Choi, N. W., Clarke, E. A., Coleman, M. P., Hakama, M., Koch, M., Langmark, F., Neal, F. E., Pettersson, F., Pompke-Kirn, V., Prior, P., and Storm, H. H. (1987). Second malignancy following testicular cancer, ovarian cancer, and Hodgkin's disease: an international collaborative study among cancer registries. *Int. J. Cancer* **39**, 571-85.
Karnofsky, D. A., Abelmann, W. H., Craver, L. F., and Burchenal, J. H. (1948). The use of nitrogen mustard in the palliative treatment of carcinoma: with particular reference to bronchogenic carcinoma. *Cancer* **1**, 634-56.
Karpas, C. M. and Jawahiri, K. (1964). Testicular embryonal carcinomas with adult teratomatous metastases. *J. Urol.* **91**, 387-91.
Kennedy, K. A. (1987). Hypoxic cells as specific targets for chemotherapy. *Anti-Cancer Drug Design* **2**, 181-94.
Kiang, D. T., Gay, J., Goldman, A., and Kenedy, B. J. (1985). A randomized trial of chemotherapy and hormonal therapy in advanced breast cancer. *New Engl. J. Med.* **313**, 7, 1241-6.
Kingston, J. E., Hawkins, M. M., Draper, D. J., Marsden, H. B., and Kinnear Wilson, L. M. (1987). Patterns of multiple primary tumours in patients treated for cancer during childhood. *Br. J. Cancer* **56**, 331-8.
Klein, G. (1987). The approaching era of the tumor suppressor genes. *Science* **238**, 1539-45.

Klein, G. and Klein, E. (1985). Evolution of tumours and the impact of molecular genetics. *Nature* **315**, 190–5.

Klein, G. and Klein, E. (1986). Conditioned tumorigenicity of activated oncogenes. *Cancer Res.* **46**, 3211–24.

Klein, G., Friberg, S., Wiener, F., and Harris, H. (1973). Hybrid cells derived from fusion of TA3-Ha ascites carcinoma with normal fibroblasts. I. Malignancy, karyotype and formation of isoantigenic variants. *J. Cancer Inst.* **50**, 1259–68.

Klein, H. O. (1972). Synchronization of tumor cell proliferation and the timing of cytostatic drugs. *Rev. Europ. Etudes Clin. et Biol.* **17**, 835–8.

Knudson, A. G. (1971). Mutation and cancer: Statistical study of retinoblastoma. *Proc. Natl. Acad. Sci. USA* **68**, 820–3.

Knudson, A. G. (1973). Mutation and human cancer. *Adv. Cancer Res.* **17**, 317–52.

Koprowski, H., Herlyn, D., Lubeck, M., De Freitas, E., and Sears, H. F. (1984). Human anti-idiotype antibodies in cancer patients: is the modification of the immune response beneficial for the patient? *Proc. Natl. Acad. Sci. USA* **81**, 216–19.

Korycka, B. M. and Hill, R. P. (1989). Dynamic heterogeneity: experimental metastasis studies with RIF-1 fibrosarcoma. *Clin. Expt. Metast.* **7**, 107–16.

Koufos, A., Hansen, M. F., Lampkin, B. C., Workman, M. L., Copeland, N. G., Jenkins, N. A. and Cavanee, W. K. (1984). Loss of alleles at loci on human chromosome 11 during genesis of Wilms' tumour. *Nature* **309**, 170–3.

Krowicki, K., Balzarini, J., De Clercq, E., Newman, R. A., and Lown, J. W. (1988). Novel DNA groove binding alkylators: Design, synthesis and biological evaluation. *J. Med. Chem.* **31**, 341–5.

Kushner, B. H., Kwon, J-H., Gulati, S. C., and Castro-Malaspina, H. (1987). Preclinical assessment of purging with VP-16-213: key role for long-term marrow cultures. *Blood* **69**, 65–71.

Lafreniere, R. and Rosenberg, S. A. (1985a). Successful immunotherapy of murine experimental hepatic metastases with lymphokine-activated killer cells and recombinant interleukin 2. *Cancer Res.* **45**, 3735–41.

Lafreniere, R. and Rosenberg, S. A. (1985b). Adoptive immunotherapy of murine hepatic metastases with lymphokine-activated killer (LAK) cells and recombinant interleukin 2 (RIL2) can mediate the regression of both immunogenic and nonimmunogenic sarcomas and an adenocarcinoma. *J. Immunol.* **135**, 4273–80.

Land, H., Parada, L. F., and Weinberg, R. A. (1983). Tumorigenic conversion of primary embryo fibroblasts requires at least two cooperating oncogenes. *Nature* **304**, 596–602.

Langdon, W. Y., Harris, A. W., Cory, S., and Adams, J. M. (1986). The c-*myc* oncogene perturbs B lymphocyte development in Eu-*myc* transgenic mice. *Cell* **47**, 11–18.

Lee, W-H., Bookstein, R., Hong, F., Young, L-J., Shew, Y-Y., and Lee, E. Y-H. P. (1987). Human retinoblastoma susceptibility gene: Cloning, identification and sequence. *Science* **235**, 1394–9.

Leighton, J. (1965). Inherent malignancy of cancer cells possibly limited by genetically differing cells in the same tumour. *Acta Cytol.* **9**, 138–40.

Le Serve, A. W. and Hellmann, K. (1972). Metastasis and the normalization of tumour blood vessels by ICRF159: a new type of drug action. *Br. Med. J.* **1**, 597–601.

Levy, R. and Miller, R. A. (1983). Tumor therapy with monoclonal antibodies. *Fed. Proc.* **42**, 2650–6.

References

Ling, V., Chambers, A. F., Harris, J. F., and Hill, R. P. (1985). Quantitative genetic analysis of tumour progression. *Cancer Metastasis Rev.* **4**, 173-94.

Link, M. P., Goorin, A. M., Miser, A. W., Green, A. A., Pratt, C. B., Belasco, J. B., Pritchard, J., Malpas, J. S., Baker, A. R., Kirkpatrick, J. A., Ayala, A. G., Shuster, J. J., Abelson, H. T., Simone, J. V., and Vietti, T. J. (1986). The effect of adjuvant chemotherapy on relapse-free survival in patients with osteosarcoma of the extremity. *New Engl. J. Med.* **314**, 1600-6.

Liotta, L. A., Kleinerman, J., and Saidel, G. M. (1974). Quantitative relationship of intravascular tumor cells, tumor vessels and pulmonary metastases following tumor implantation. *Cancer Res.* **34**, 997-1004.

Liotta, L. A., Kleinerman, J., and Saidel, G. M. (1976). The significance of hematogenous tumor cell clumps in the metastatic process. *Cancer Res.* **34**, 889-94.

Liotta, L. A., Wewer, U., Rao, N. C., Schiffmann, E., Stacke, M., Guirguis, R., Thorgeirsson, U., Muschel, R., and Sobel, M. (1987). Biochemical mechanisms of tumor invasion and metastasis. *Anti-Cancer Drug Design* **2**, 195-202.

Lippman, M. E. (ed.) (1986). *Proceedings of the NIH consensus development conference on adjuvant chemotherapy and endocrine therapy for breast cancer.* National Cancer Institute Monographs, No. 1.

Lippman, M. E. and Bolan, G. (1975). Oestrogen-responsive human breast cancer line in long term tissue culture. *Nature* **256**, 592-3.

Lippman, M. E. and Chabner, B. A. (1986). Editorial overview. *Natl. Cancer Inst. Monographs* **1**, 5-9.

Little, C. D., Nau, M. M., Carney, D. N., Gazdar, A. F., and Minna, J. D. (1983). Amplification and expression of the c-*myc* oncogene in human lung cancer cell lines. *Nature* **306**, 194-6.

Little, J. M. (1987). A method of calculating the value of palliative care of cancer patients. *Aust. N.Z. J. Surg.* **57**, 393-7.

Lotze, M. T., Grimm, E. A., Mazumder, A., Strausser, J. L., and Rosenberg, S. A. (1981). Lysis of fresh and cultured autologous tumor by human lymphocytes cultured in T-cell growth factor. *Cancer Res.* **41**, 4420-5.

Low, C. M. L., Drew, H. R., and Waring, M. J. (1984). Sequence-specific binding of echinomycin to DNA: evidence for conformational changes affecting flanking sequences. *Nucl. Acids Res.* **12**, 4865-79.

Lown, J. W., Krowicki, K., Balzarini, J., and De Clercq, E. (1986). Structure-activity relationship of novel oligopeptide antiviral and antitumour agents related to netropsin and distamycin. *J. Med. Chem.* **29**, 1210-14.

Luria, S. E. and Delbruck, M. (1943). Mutations of bacteria from virus sensitivity to virus resistance. *Genetics* **28**, 491-511.

Lyon, M. F. (1974). Review lecture. Mechanisms and evolutionary origins of variable X-chromosome activity in mammals. *Proc. R. Soc. London. B* **187**, 243-68.

McCarty, J. S. (Jr), Szabo, E., Flowers, J. L., Cox, E. B., Leight, J. S., Miller, L., Konrath, J., Soper, J. T., Budwit, D. A., Creasman, W. T., Siegler, H. F., and McCarty, J. S. (Sr). (1986). Use of a monoclonal anti-estrogen receptor antibody in the immunohistochemical evaluation of human tumors. *Cancer Res.* (Suppl.) **46**, 4244-8.

McGuire, W. L., Clark, G. M., Dressler, L. G., and Owens, M. A. (1986). Role of steroid hormone receptors as prognostic factors in primary breast cancer. *National Cancer Inst. Monographs* **1**, 19–23.

McLaren, A. L. (1975). *Mammalian chimaeras*. Cambridge University Press, Cambridge and London.

Maclean, N. and Hall, B. K. (1987). *Cell commitment and differentiation*. Cambridge University Press, Cambridge and London.

McLeod, G. R., Beardmore, G. L., Little, J. H., Quinn, R. L., and Davis, N. C. (1971). Results of treatment of 361 patients with malignant melanoma in Queensland. *Med. J. Australia* **1**, 1211–16.

Marshall, C. J. and Rigby, P. W. (1984). Viral and cellular genes involved in oncogenesis. *Cancer Surveys* **3**, 183–214.

Mathe, G. (1977). *Dossier cancer*, p. 292. Stock, Paris.

Mathew, C. G. P., Chin, K. S., Easton, D. F., Thorpe, K., Carter, C., Liou, G. I., Fong, S.-L., Bridges, C. D. B., Haak, H., Kruseman, A. C. N., Schifter, S., Hansen, H. H., Telenius, H., Telenius-Berg, M., and Ponder, B. A. J. (1987). A linked genetic marker for multiple endocrine neoplasia type 2A on chromosome 10. *Nature* **328**, 527–8.

Mazumder, A., Eberlein, T. J., Grimm, E. A., Wilson, D. J., Keenan, A. M., Aamodt, R., and Rosenberg, S. A. (1984). Phase I study of the adoptive immunotherapy of human cancer with lectin activated autologous mononuclear cells. *Cancer* **53**, 896–905.

Medical Research Council Working Party On Bone Sarcoma (1986). A trial of chemotherapy in patients with osteosarcoma. *Br. J. Cancer* **53**, 513–18.

Melicow, M. M. (1982). The three steps to cancer; a new concept of carcinogenesis. *J. Theor. Biol.* **44**, 471–511.

Metcalfe, D. (1987). The Wellcome Foundation Lecture, 1986. The molecular control of normal and leukaemic granulocytes and macrophages. *Proc. R. Soc. Lond. B* **230**, 389–423.

Micklem, H. S. (1986). Clonal succession: a commentary. *Blood Cells* **12**, 119–26.

Migliori, R. J., Gruber, S. A., Sawyer, M. D., Hoffman, R., Ochoa, A., Bach, F. H., and Simmons, R. L. (1987). Lymphokine-activated killer (LAK) cells can be focused at sites of tumor growth by products of macrophage activation. *Surgery* **102**, 155–62.

Miller, P. S. and Ts'o, P. O. P. (1987). A new approach to chemotherapy based on molecular biology and nucleic acid chemistry: matagen (masking tape for gene expression). *Anti-Cancer Drug Design* **2**, 117–28.

Miller, P. S., McParland, K. B., Jayaraman, K., and Ts'o, P. O. P. (1981). Biochemical and biological effects of nonionic nucleic acid methylphosphonates. *Biochemistry* **20**, 1874–80.

Mintz, B. and Illmensee, K. (1975). Normal genetically mosaic mice produced from malignant teratocarcinoma cells. *Proc. Natl. Acad. Sci. USA* **72**, 3585–9.

Mitchell, J. S. (1971). *Cancer. If curable, why not cured?* Heffer, Cambridge.

Moller, G. (ed.) (1982). Antibody carriers of drugs and toxins in tumor therapy. *Immunol. Revs.* **62**, 1–216.

Monier, R. (1987). More about growth factors and oncogenes. *The Cancer Journal* **1**, 393–4.

Mule, J. J., Shu, S., Schwarz, S. L., and Rosenberg, S. A. (1984). Adoptive immunotherapy of established pulmonary metastases with LAK cells and recombinant interleukin 2. *Science* **225**, 1487-9.

Mule, J. J., Shu, S., and Rosenberg, S. A. (1985). The anti-tumor efficacy of lymphokine-activated killer cells and recombinant interleukin 2 *in vivo*. *J. Immunol.* **135**, 646-52.

Mule, J. J., Smith, C. A., and Rosenberg, S. A. (1987). Interleukin 4 (B cell stimulatory factor 1) can mediate the induction of lymphokine-activated killer cell activity directed against fresh tumor cells. *J. Exper. Med.* **166**, 792-7.

Mulvihill, J. J., Myers, M. H., Connelly, R. R., Byrne, J., Austin, D. F., Bragg, K., Cook, J. W., Hassinger, D. D., Holmes, F. F., Holmes, G. F., Krauss, M. R., Latourette, H. B., Meigs, J. W., Naughton, M. D., Steinhorn, S. C., Teta, M. J., and Weyer, P. J. (1987). Cancer in long-term survivors of childhood and adolescent cancer. *Lancet* **2**, 813-17.

Murphree, A. L. and Benedict, W. F. (1984). Retinoblastoma: Clues to human oncogenesis. *Science* **223**, 1028-33.

Murphy, P., Alexander, P., Senior, P. V., Fleming, J., Kirkham, N., and Taylor, I. (1988). Mechanisms of organ selective tumour growth by bloodborne cancer cells. *Br. J. Cancer* **57**, 19-31.

Muul, L. M., Spiess, P. J., Director, E. P., and Rosenberg, S. A. (1987). Identification of specific cytolytic immune responses against autologous tumor in humans bearing malignant melanoma. *J. Immunol.* **138**, 989-95.

National Cancer Institute Monographs (1976). *Conference on Spontaneous Regression of Cancer*, Vol. 44.

Nau, M. N., Carney, D. N., Battey, J., Johnson, B., Little, C., Gazdar, A., and Minna, J. D. (1984). Amplification, expression and rearrangement of c-*myc* and N-*myc* oncogenes in human lung cancer. *Curr. Top. Microbiol. Immunol.* **113**, 172-7.

Naylor, S. L., Johnson, B. E., Minna, J. D., and Sakaguchi, A. Y. (1987). Loss of heterozygosity of chromosome 3p marker in small cell lung cancer. *Nature* **329**, 451-4.

Neuberger, M. S., Williams, G. T., and Fox, R. O. (1984). Recombinant antibodies possessing novel effector functions. *Nature* **312**, 604-8.

Nissen-Meyer, R. (1982). The Scandinavian clinical trials. *Experientia* **41**, 571-9.

Noda, M., Ko, M., Ogura, A., Liu, D-G., Amano, T., Takano, T., and Ikawa, Y. (1985). Sarcoma viruses carrying *ras* oncogenes induce differentiation-associated properties in a neuronal cell line. *Nature* **318**, 73-5.

Nolvadex Adjuvant Trial Organisation (1985). Controlled trial of tamoxifen as single adjuvant agent in management of early breast cancer. *Lancet* **1**, 836-43.

Nowell, P. C. (1976). The clonal evolution of tumor cell populations. *Science* **194**, 23-8.

Nowell, P. C. (1982). Genetic instability in cancer cells. In *Tumor cell heterogeneity: origins and implications* (ed. A. H. Owens, D. S. Coffey, and S. B. Baylin), pp. 351-65. Academic Press, New York.

Oldham, R. K. (ed.) (1987). *Principles of cancer biotherapy*. Raven Press, New York.

Padmanabhan, N., Howell, A., and Rubens, R. D. (1986). Mechanism of action of adjuvant chemotherapy in early breast cancer. *Lancet* **2**, 411-14.

Paget, S. (1889). The distribution of secondary growth in the breast. *Lancest* **1**, 571.

Papac, R. J., Weinberg, J. B., Son, Y. H., Sasaki, C., Fisher, D. B., Lawrence, R., Rockwell, S., Sartorelli, A. C., and Fisher, J. J. (1987). Prospective randomized trial of radiation therapy (RT) +/− mitomycin C (MC) in head and neck cancer. *Proc. Amer. Soc. Clin. Oncol.* **6**, 126.

Pappaioannou, V. E., McBurney, M. W., Gardner, R. L., and Evans, R. L. (1975). Fate of teratocarcinoma cells injected into early mouse embryos. *Nature* **258**, 70-3.

Paul, J. (1988). The role of oncogenes in carcinogenesis. In *Theories of carcinogenesis* (ed. O. H. Iverson), pp. 45-60. Hemisphere, Washington, DC.

Penn, I. (1970). *Malignant tumors in organ transplant recipients.* Springer Verlag, Berlin and New York.

Peto, R. (1977). Epidemiology, multistage models and short mutagenicity tests. In *Origins of human cancer* (ed. H. H. Hiatt, J. D. Watson, and J. A. Winsten), pp. 1403-28. Cold Spring Harbor Laboratory, New York.

Peto, R., Pike, M. C., Armitage, P., Breslow, N. E., Cox, D. R., Howard, S. V., Mantel, N., McPherson, K., Peto, J., and Smith, P. G. (1976). Design and analysis of randomized clinical trials requiring prolonged observation of each patient. I. Introduction and design. *Br. J. Cancer* **34**, 585-612.

Peto, R., Pike, M. C., Armitage, P., Breslow, N. E., Cox, D. R., Howard, S. V., Mantel, N., McPherson, K. Peto, J., and Smith, P. G. (1977). Design and analysis of randomized observation of each patient. II. Analysis and examples. *Br. J. Cancer* **35**, 1-39.

Pierce, G. B. and Speers, W. C. (1988). Tumors as caricatures of the process of tissue renewal: prospect for therapy by directing differentiation. *Cancer Res.* **48**, 1996-2004.

Pierce, G. B. and Wallace, C. (1971). Differentiation of malignant to benign cells. *Cancer Res.* **31**, 127-34.

Pimm, M. V., Embleton, M. J., and Baldwin, R. W. (1980). Multiple antigenic specificities within primary 3-methylcholanthrene-induced rat sarcomas and metastases. *Int. J. Cancer* **25**, 621-62.

Pohl, J., Radler-Pohl, A., Heicappel, R., and Schirrmacher, V. (1988). Oncogene expression in related cancer lines differing in metastatic capacity. *Clin. Expl. Metastasis* **6**, 201-11.

Poste, G., Doll, J., and Fidler, I. J. (1981). Interactions among clonal subpopulations affect stability of the metastatic phenotype in polyclonal populations of B16 melanoma cells. *Proc. Natl. Acad. Sci. USA* **78**, 6226-30.

Poste, G., Tzeng, J., Doll, J., Greig, R., Rieman, D., and Zeidman, I. (1982). Evolution of tumor cell heterogeneity during progressive growth of individual lung metastases. *Proc. Natl. Acad. Sci. USA* **79**, 6574-8.

Prehn, R. T. (1970). Analysis of antigenic heterogeneity within individual 3-methylcholanthrene-induced mouse sarcomas. *J. Natl. Cancer Inst.* **45**, 1039-45.

Prehn, R. T. and Main, J. M. (1957). Immunity to methylcholanthrene-induced sarcomas. *J. Natl. Cancer Inst.* **18**, 769-78.

Price, J. E., Carr, D., and Tarin, D. (1984). Spontaneous and induced metastasis of naturally occurring tumors in mice: analysis of cell shedding into the blood. *J. Natl. Cancer Inst.* **73**, 1319-26.

Radler-Pohl, A., Pohl, J., and Schirrmacher, V. (1988). Selective enhancement of metastatic capacity in mouse bladder carcinoma cells after transfection of DNA from liver metastases of human colon carcinoma. *Int. J. Cancer* **41**, 840-6.

Ragaz, J., Baird, R., Rebbeck, P., Goldie, J., Coldman, A., and Spinelli, J. (1985). Neoadjuvant (preoperative) chemotherapy for breast cancer. *Cancer* **56**, 719–24.

Raskind, W. H. and Fialkow, P. J. (1987). The use of cell markers in the study of human hematopoietic neoplasia. *Adv. Cancer Res.* **49**, 127–67.

Reddy, A. L. and Fialkow, P. J. (1983). Papillomas induced by initiation-promotion differ from those induced by carcinogen alone. *Nature* **304**, 69–71.

Reichmann, L., Clark, M., Waldmann, H., and Winter, G. (1988). Reshaping human antibodies for therapy. *Nature* **332**, 323–7.

Ribeiro, G. and Swindells, R. (1985). The Christie Hospital tamoxifen (Nolvadex) adjuvant trial for operable breast carcinoma—7 year results. *Eur. J. Cancer Clin. Oncol.* **21**, 897–900.

Ritz, J., Pesando, J. M., Sallan, S. E., Clavell, L. A., Notis-McConarty, J., Rosenthal, P., and Schlossman, S. F. (1981). Serotherapy of acute lymphoblastic leukemia with monoclonal antibody. *Blood* **58**, 141–52.

Roitt, I. M., Brostoff, J., and Male, D. K. (1985). *Immunology*. Churchill Livingstone, Edinburgh.

Rosenau, W. and Morton, D. L. (1966). Tumour-specific inhibition of growth of methylcholanthrene-induced sarcomas *in vivo* and *in vitro* by sensitized isologous lymphoid cells. *J. Natl. Cancer Inst.* **36**, 825–36.

Rosenberg, S. A., Mule, J. J., Spiess, P. J., Reichert, C. M., and Schwarz, S. L. (1985a). Regression of established pulmonary metastases and subcutaneous tumor mediated by the systemic administration of high-dose recombinant interleukin 2. *J. Exper. Med.* **161**, 1169–88.

Rosenberg, S. A., Lotze, M., Muul, L. M., Leitman, S., Chay, A. E., Ettinghausen, S. E., Matory, Y. L., Skibber, J. M., Shiloni, E., Vetto, J. T., Seipp, C. A., Simpson, C., and Reichert, C. M. (1985b). Observations on the systemic administration of autologous lymphokine-activated killer cells and recombinant interleukin-2 to patients with metastatic cancer. *New Engl. J. Med.* **313**, 1485–92.

Rosenberg, S. A., Spiess, P., and Lafreniere, R. (1986). A new approach to the adoptive immunotherapy of cancer with tumor-infiltrating lymphocytes. *Science* **233**, 1318–21.

Rosenberg, S. A., Lotze, M. T., Muul, L. M., Chang, A. E., Avis, F. P., Leitman, S., Linehan, W. M., Robertson, C. N., Lee, R. E., Rubin, J. T., Seipp, C. A., Simpson, C. G., and White, D. E. (1987). A progress report on the treatment of 157 patients with advanced cancer using lymphokine-activated killer cells and interleukin-2 or high-dose interleukin-2 alone. *New Engl. J. Med.* **316**, 889–97.

Rosenstein, M., Yron, K., Kaufmann, Y., and Rosenberg, S. A. (1984). Lymphokine-activated killer cells: lysis of fresh syngeneic natural killer-resistant murine tumor cells by lymphocytes cultured in interleukin 2. *Cancer Res.* **44**, 1946–53.

Rouleau, G. A., Werterlecki, W., Haines, J. L., Hobbs, W. J., Trofatter, J. A., Seizeninger, B. D., Martuza, R. L., Superneau, D. W., Conneally, P. M., and Gusella, J. F. (1987). Genetic linkage of bilateral acoustic neurofibromatosis to a DNA marker on chromosome 22. *Nature* **329**, 246–8.

Rous, P. and Beard, J. W. (1935). The progression to carcinoma of virus-induced rabbit papillomas (Shope). *J. Exper. Med.* **62**, 523–48.

Rowland, G. F., O'Neill, G. J., and Davis, D. A. L. (1975). Suppression of tumour growth in mice by a drug-antibody conjugate using a novel approach to linkage. *Nature* **255**, 487-8.

Rubin, H. (1984). Mutations and oncogenes—cause or effect. *Nature* **309**, 518.

Sachs, L. (1974). Regulation of membrane changes, differentiation and malignancy in carcinogenesis. *The Harvey Lectures, 1972-1973, Series 68*, pp. 1-35.

Sachs, L. (1984). Normal regulators, oncogenes and the reversibility of malignancy. *Cancer Surveys* **3**, 210-28.

Sachs, L. (1987*a*). The Wellcome Foundation Lecture, 1986. The molecular regulators of normal and leukaemic blood cells. *Proc. R. Soc. B* **231**, 289-312.

Sachs, L. (1987*b*). Induction of cell differentiation and bypassing of genetic defects in cancer therapy. In *New avenues in developmental cancer chemotherapy* (ed. K. R. Harrap and T. A. Connors), pp. 187-204. Academic Press, London.

Sager, R. (1985). Genetic suppression of tumor formation. *Adv. Cancer Res.* **44**, 43-68.

Sakamoto, N. and Tanaka, N. G. (1988). Mechanism of the synergistic effect of heparin and cortisone against angiogenesis and tumour growth. *The Cancer Journal* **2**, 9-13.

Sartorelli, A. C. (1988). Therapeutic attack of hypoxic cells of solid tumors: presidential address. *Cancer Res.* **48**, 775-8.

Sartorelli, A. C., Morin, M. J., and Ishiguro, K. (1987). Cancer chemotherapeutic agents as inducers of leukemia cell differentiation. In *New avenues in developmental chemotherapy* (ed. K. R. Harrap and T. A. Connors), pp. 229-44. Academic Press, London.

Sattaur, O. (1984). Cancer genes—the enemy within. *New Scientist* **104**, 12-16.

Saxon, P. J., Srivatsan, E. S., and Stanbridge, E. J. (1986). Introduction of human chromosome 11 via microcell transfer controls tumorigenic expression of HeLa cells. *EMBO J.* **5**, 3461-6.

Schabel, F. M. (1969). The use of tumor growth kinetics in planning 'curative' chemotherapy of advanced solid tumors. *Cancer Res.* **29**, 2384-9.

Schabel, F. M. (1975). Concepts for systemic treatment of micrometastases. *Cancer* **35**, 15-24.

Schirrmacher, V. (1985). Cancer metastasis. *Adv. Cancer Res.* **43**, 1-73.

Schirrmacher, V., Bosslet, K., Altevogt, P., Russmann, E., Beck, L., and Fogel, M. (1982). Cited in Schirrmacher (1985).

Schwab, M., Ellison, J., Busch, M., Rosenau, M., Varmus, H. E., and Bishop, J. M. (1984). Enhanced expression of the human gene N-*myc* consequent to amplification of DNA may contribute to malignant progression of neuroblastoma. *Proc. Natl. Acad. Sci. USA* **81**, 4940-4.

Scottish Breast Cancer Trials Committee (1987). Adjuvant tamoxifen in the management of operable breast cancer: the Scottish trial. *Lancet* **2**, 171-5.

Seizinger, B. R., Martuza, R. L., and Gusella, J. F. (1986). Loss of genes on chromosome 22 in tumorigenesis of human acoustic neuroma. *Nature* **322**, 644-7.

Seizinger, B. R., Rouleau, G. A., Ozelius, L. J., Lane, A. H., Faryniarz, A. G., Chao, M. V., Huson, S., Korf, B. R., Parry, D. M., Pericak-Vance, A., Collins, F. S., Hobbs, W. J., Falcone, B. G., Ianazzi, J. A., Roy, J. C., St George-Hyslop, P. H., Tanzi, R. E., Bothwell, M. A., Upadhyaya, M., Harper, P., Goldstein, A. E., Hoover, D. L., Bader, J. L., Spence, M. A., Mulvihill, J. J., Aylsworth, A. S., Vance, J. M., Rossenwasser, G. O. D., Gaskell, P. C., Roses, A. D., Martuza,

R. L., Breakefield, X. O., and Gusella, J. F. (1987). Genetic linkage of von Recklinghausen neurofibromatosis to the nerve growth factor receptor gene. *Cell* **49**, 589-94.

Shapiro, S., Venet, W., Strax, P., Venet, L., and Roeser, R. (1982). Ten to fourteen year effect of screening on breast cancer. (1982). *J. Natl. Cancer Inst.* **69**, 349-55.

Shivas, A. A. and Finlayson, N. D. C. (1965). The resistance of arteries to tumour invasion. *Br. J. Cancer* **19**, 486-9.

Simon-Assmann, P., Bouziges, P., Haffen, K., and Kedinger, M. (1988). Extracellular matrix in intestinal morphogenesis and differentiation. *Br. J. Cancer* **58**, 223 (abst.).

Simpson, N. E., Kidd, K. K., Goodfellow, P. J., McDermid, H., Myers, S., Kidd, J. R., Jackson, C. E., Duncan, A. M. V., Farrer, L. A., Brasch, K., Castiglione, C., Genel, M., Gertner, J., Greenberg, C. R., Gusella, J. F., Holden, J. J. A., and White, B. N. (1987). Assignment of multiple endocrine neoplasia type 2A to chromosome 10 by linkage. *Nature* **328**, 528-30.

Siracky, J. (1979). An approach to the problem of heterogeneity of human tumour-cell populations. *Br. J. Cancer* **39**, 570-7.

Skipper, H. E. (1983). The forty-year-old mutation theory of Luria and Delbruck and its pertinence to cancer chemotherapy. *Adv. Cancer Res.* **40**, 331-63.

Skipper, H. E. (1985). *Critical variables in the design of combination chemotherapy regimens to be used alone or in an adjuvant setting.* Booklet No. 8. Southern Research Institute, Birmingham, Alabama.

Skipper, H. E. and Schabel, F. M. (1984). Tumor cell heterogeneity: implications with respect to classification of cancers by chemotherapeutic effect. *Cancer Treat. Rep.* **68**, 43-61.

Skipper, H. E., Schabel, F. M., and Wilcox, W. S. (1964). Experimental evaluation of potential anti-cancer agents. XIII. On the criteria and kinetics associated with 'curability' of experimental leukemia. *Cancer Chemoth. Rep.* **35**, 1-111.

Skipper, H. E., Schabel, F. M., and Wilcox, W. S. (1965). Experimental evaluation of potential anti-cancer agents. XIV. Further study of certain basic concepts underlying chemotherapy of leukemia. *Cancer Chemoth. Rep.* **45**, 5-28.

Sklar, J., Clearly, M. L., Thielmans, K., Gralow, J., Warnke, R., and Levy, R. (1984). Bioclonal B-cell lymphoma. *New Engl. J. Med.* **311**, 20-7.

Smithers, D. W. (1969). Maturation in human tumours. *Lancet* **2**, 949-52.

Smyth, J. F. (1980). Science and serendipity in cancer medicine. *University of Edinburgh Inaugural Lecture, No. 69.*

Solomon, E., Voss, R., Hall, V., Bodmer, W. F., Jass, J. R., Jeffreys, A. J., Lucibello, F. C., Pate, I., and Rider, S. H. (1987). Chromosome 5 allele loss in human colorectal carcinomas. *Nature* **328**, 616-19.

Spandidos, D. A. and Wilkie, N. M. (1984). Malignant transformation of early passage rodent cells by a single mutated human oncogene. *Nature* **310**, 469-75.

Sparkes, R. S., Murphree, A. L., Lingua, R. W., Sparkes, M. C., Field, L. L., Funderburk, S. J., and Benedict, W. F. (1983). Gene for hereditary retinoblastoma assigned to human chromosome 13 by linkage to esterase D. *Science* **219**, 971-3.

Speers, W. C. (1982). Conversion of malignant murine embryonal carcinoma to benign teratomas by chemical induction of differentiation *in vivo*. *Cancer Res.* **42**, 1843-7.

Sporn, M. B. and Todaro, G. J. (1980). Autocrine secretion and malignant transformation of cells. *New Engl. J. Med.* **303**, 878-80.

Standbridge, E. J. (1987). Genetic regulation of tumorigenic expression in somatic cell hybrids. *Adv. Viral Oncol.* **6**, 83–101.

Stern, D. F., Hare, D. L., Cecchini, M. A., and Weinberg, R. A. (1987). Construction of a novel oncogene based on synthetic sequences encoding epidermal growth factor. *Science* **235**, 321–4.

Steward, W. P., Scarffe, J. H., Borkett, K., Bonnem, E., and Crowther, D. (1989). High-dose melphalan with granulocyte–macrophage colony-stimulating factor (GM-CSF) in the treatment of metastatic colo-rectal carcinoma (abst.). *Br. J. Cancer* **60**, 449.

Stewart, T. A. and Mintz, B. (1981). Successive generations of mice produced from an established culture of euploid, teratocarcinoma cells. *Proc. Natl. Acad. Sci. USA* **78**, 6314–18.

Stoker, M. G. P. (1972). Tumour viruses and the sociology of fibroblasts. *Proc. R. Soc. Lond. B* **181**, 1–17.

Storb, R. (1987). Critical issues in bone marrow transplantation. *Transplantation Proc.* **19**, 2774–81.

Sumegi, J., Hedberg, T., Bjorkholm, M., Godal, T., Mellstedt, H., Nilsson, M., Perlmann, C., and Klein, G. (1985). Amplification of the c-*myc* oncogene in human plasma cell leukemia. *Int. J. Cancer* **36**, 367–71.

Sutherland, C. M. (1986). Regional isolation perfusion for malignant melanoma of the extremities. *Cancer Topics* **5**, 114–15.

Tabar, L., Fagerberg, C. J. G., Gad, A., Baldetorp, L., Holmberg, L. H., Grontoft, O., Ljungquist, V., Lundstrom, B., and Manson, J. C. (1985). Reduction in mortality from breast cancer after mass screening with mammography. *Lancet* **1**, 829–32.

Talmadge, J. E. and Fidler, L. J. (1982). Cancer metastasis is selective or random depending on the parent population. *Nature* **297**, 593–4.

Talmadge, J. E., Wolman, S. R., and Fidler, L. J. (1982). Evidence for the clonal origin of spontaneous metastases. *Science* **217**, 361–3.

Talmadge, J. E., Benedict, K., Madsen, J., and Fidler, I. J. (1984). Development of biological diversity and suceptibility to chemotherapy in murine cancer metastases. *Cancer Res.* **44**, 3801–5.

Tarin, D. (1985). Clinical and experimental studies on the biology of metastasis. *Biochim. Biophys. Acta* **780**, 227–35.

Taylor, I., Machin, D., Mullee, M., Trotter, G., Cooke, T., and West, C. (1985). A randomized controlled trial of adjuvant portal vein cytotoxic perfusion in colorectal cancer. *Brit. J. Surg.* **72**, 359–63.

Thatcher, N. (1989). The role of growth factors in patient management (abst.). *Br. J. Cancer* **60**, 442.

Thiele, C. J., Reynolds, C. P., and Israel, M. A. (1985). Decreased expressed of N-*myc* precedes retinoic acid-induced morphological differentiation of human neuroblastoma. *Nature* **313**, 404–6.

Thomas, E. D. (1984). Current status of bone marrow transplantation. In *Current status of clinical organ transplantation* (ed. G. M. Abouna), pp. 269–83, Martininus Nijhoff, Boston.

Thomas, E. D., Storb, R., Clift, R. A., Fefer, A., Johnson, F. L., Neiman, P. E.,

Lerner, K. G., Glucksberg, H., and Buckner, C. D. (1975). Bone-marrow transplantation. *N. Engl. J. Med.* **292**, 832-43, 895-902.

Thorgeirsson, U. P., Turpeennemi-Hujanen, T., Williams, J. E., Westin, E. H., Heilman, C. A., Talmadge, J. E., and Liotta, L. A. (1985). NIH/3T3 cells transfected with human tumor DNA containing activated *ras* oncogenes express the metastatic phenotype in nude mice. *Molec. Cell. Biol.* **5**, 259-62.

Thorpe, P. E. and Ross, W. C. J. (1982). The preparation and cytotoxic properties of antibody-toxin conjugates. *Immunol. Revs.* **62**, 119-58.

Thuong, N. T., Asseline, V., Roig, V., Takasugi, M., and Helene, C. (1987). Oligo(α-deoxynucleotide)s covalently linked to intercalating agents: Differential binding to ribo- and deoxyribonucleotides and stability towards nuclease digestion. *Proc. Natl. Acad. Sci. USA* **84**, 5129-33.

Todaro, G. (1978). RNA-tumour-virus genes and transforming genes. Patterns of transmission. *Br. J. Cancer* **37**, 139-58.

Todaro, G. J., Marquardt, H., Twardzik, D. R., Johnson, P. A., Fryling, C. M., and De Larco, J. E. (1982). Transforming growth factors produced by tumor cells. In *Tumor cell heterogeneity: origins and implications* (ed. A. H. Owens, D. S. Coffey, and S. B. Baylin), pp. 205-24. Academic Press, New York.

Tonegawa, S. (1983). Somatic generation of antibody diversity. *Nature* **302**, 575-81.

Toulmé, J. J., Krisch, H. M., Loreau, N., Thuong, N. T., and Helene, C. (1986). Specific inhibition of mRNA translation by complementary ologonucleotides covalently linked intercalating agents. *Proc. Natl. Acad. Sci. USA* **83**, 1227-31.

Vaage, J. (1988). Metastasizing potentials of mouse mammary tumors and their metastases. *Int. J. Cancer* **41**, 855-8.

Vaage, J. (1989). Loss of metastasizing potential in mouse mammary tumors. *Clin. Expl. Metastasis* **7**, 373-8.

Vande Woude, G. F., Oskarsson, M., McGeady, M. L., Seth, A., Propst, F., Schmidt, M., Paules, R., and Blair, D. G. (1987). Sequences that influence the transforming activity and expression of the *mos* oncogene. *Adv. Viral Oncol.* **6**, 71-81.

Volpe, J. P. and Milas, L. (1988). Metastatic instability of murine tumor metastases: dependent on tumor type. *Clin. Expl. Metastasis* **6**, 333-46.

Vousden, K. H., Eccles, S. A., Purvies, H., and Marshall, C. J. (1986). Enhanced spontaneous metastasis of mouse carcinoma cells transfected with an activated c-Ha-*ras*-1 gene. *Int. J. Cancer* **37**, 425-33.

Webb, C. W., Gootwine, E., and Sachs, L. (1984). Developmental potential of myeloid leukemia cells injected into midgestation embryos. *Devel. Biol.* **101**, 221-4.

Weinberg, R. A. (1985). The action of oncogenes in the cytoplasm and nucleus. *Science* **230**, 770-6.

Weinberg, R. A. (1988). Finding the anti-oncogene. *Scientific American* **259**, No. 3, 34-41.

Weiss, L. (1980*a*). Cancer traffic from the lungs to the liver: an example of metastatic inefficiency. *Int. J. Cancer* **25**, 385-92.

Weiss, L. (1980*b*). Differences between cancer cells in primary and secondary tumours. *Pathobiol. Ann.* **10**, 51-81.

Weiss, L. and Glaves, D. (1983). Cancer cell damage at the vascular endothelium. *Ann. NY Acad. Sci.* **416**, 81-692.

Weiss, L., Ward, P. M., and Holmes, J. C. (1983a). Liver to lung traffic of cancer cells. *Int. J. Cancer* **32**, 79–83.

Weiss, L., Holmes, J. C., and Ward, P. M. (1983b). Do metastases arise from pre-existing subpopulations of cancer cells. *Br. J. Cancer* **47**, 81–9.

Weiss, R. A. (1973). Transmission of cellular genetic elements by RNA tumour viruses. In *Possible episomes in eukaryotes. IV Lepetit Colloquium.* (ed. L. Sylvestri), pp. 130–41. North Holland, Amsterdam.

Weiss, R. A., Mason, W. S., and Vogt, P. G. (1973). Genetic recombination between endogenous and exogenous avian RNA tumor viruses. *Virology* **52**, 535–52.

Weissman, B. E., Saxon, P. J., Pasquale, S. R., Jones, G. R., Geiser, A. G., and Stanbridge, E. J. (1987). Introduction of a normal human chromosome 11 into a Wilms' tumor cell line controls its tumorigenic expression. *Science* **236**, 176–80.

Wessels, N. K. (1977). *Tissue interaction and development.* Benjamin Cummings, Menlo Park, California.

West, W. H., Tauer, K. W., Yannelli, J. R., Marshall, G. D., Orr, D. W., Thurman, G. B., and Oldham, R. K. (1987). Constant-infusion recombinant interleukin-2 in adoptive immunotherapy of advanced cancer. *New Engl. J. Med.* **316**, 898–905.

Wheelock, E. F., Robinson, M. K., and Truitt, G. A. (1982). Establishment and control of the L5178Y-cell tumor dormant state in DBA/2 mice. *Cancer Metastasis Rev.* **1**, 29–44.

White, H. and Griffiths, J. D. (1976). Circulating malignant cells and fibrinolysis during resection of colorectal cancer. *Proc. R. Soc. Med.* **69**, 467–9.

Wiener, F., Fenyo, E. M., Klein, G., and Harris, H. (1972). Fusion of tumour cells with host cells. *Nature New Biology* **238**, 155–9.

Williams, G. (1988). Management of metastatic cancer of the prostate. *Prescribers' J.* **28**, 43–8.

Williams, M. (1985). Principles of radiotherapy. 1. The biological basis of radiotherapy. *Cancer Topics* **5**, 74–5.

Willis, R. A. (1952, 1973). *The spread of tumours in the human body.* 1st and 3rd edns, Butterworth, London.

Woodruff, M. F. A. (1961). New approaches to the treatment of cancer. *J. R. Coll. Surg. Edinb.* **6**, 75–92.

Woodruff, M. F. A. (1972). Residual cancer. *Harvey Lect. Series* **66**, 161–76.

Woodruff, M. F. A. (1975). Latent tumour metastases. *Nature* **258**, 776.

Woodruff, M. F. A. (1977). *On science and surgery.* Edinburgh University Press, Edinburgh.

Woodruff, M. F. A. (1980). *The interaction of cancer and host.* Grune and Stratton, New York.

Woodruff, M. F. A. (1982). Interaction of cancer and host. *Br. J. Cancer* **46**, 313–22.

Woodruff, M. F. A. (1983). Cellular heterogeneity in tumours. *Br. J. Cancer* **47**, 589–94.

Woodruff, M. F. A. (1986a). What's going on in the cancer patient? *Pathology* **18**, 175–80.

Woodruff, M. F. A. (1986b). The cytolytic and regulatory role of natural killer cells in experimental neoplasia. *Biochim. Biophys. Acta* **865**, 43–57.

Woodruff, M. F. A. (1987). The Halford Oration—the interface between medicine and

science. *Chiron* (The Journal of the University of Melbourne Medical Society). **2**, 4–10.
Woodruff, M. F. A. (1988). Tumor clonality and its biological significance. *Adv. Cancer Res.* **50**, 197–229.
Woodruff, M. F. A. and Anderson, N. A. (1963). Effect of lymphocyte depletion by thoracic duct fistula and administration of antilymphocytic serum on the survival of skin homografts in rats. *Nature* **200**, 702.
Woodruff, M. F. A. and Hodson, B. A. (1985a). The effect of passage *in vitro* and *in vivo* on the properties of murine fibrosarcomas. I. Tumorigenicity and immunogenicity. *Br. J. Cancer* **51**, 161–9.
Woodruff, M. F. A. and Hodson, B. A. (1985b). The effect of passage *in vitro* and *in vivo* on the properties of murine fibrosarcomas. II. Sensitivity to cell-mediated immunity *in vitro*. *Br. J. Cancer* **52**, 233–40.
Woodruff, M. F. A. and van Rood, J. J. (1983). Possible implications of the effect of blood transfusion on allograft survival. *Lancet* **1**, 1201–2.
Woodruff, M. F. A. and Walbaum, P. (1983). A phase-II trial of *Corynebacterium parvum* as adjuvant to surgery in the treatment of operable lung cancer. *Cancer Immunol. Immunother.* **16**, 114–16.
Woodruff, M. F. A., Nolan, B., Anderton, J. L., Abouna, G. M., Morton, J. B., and Jenkins, A. McL. (1976). Long term survival after renal transplantation in man. *Br. J. Surg.* **63**, 85–101.
Woodruff, M. F. A., Ansell, J. D., Forbes, G. M., Gordon, J. G., Burton, D. I., and Micklem, H. S. (1982). Clonal interaction in tumours. *Nature* **299**, 822–4.
Woodruff, M. F. A., Ansell, J. D., Hodson, B. A., and Micklem, H. S. (1984). Specificity of tumour associated transplantation antigens (TATA) of different clones from the same tumour. *Br. J. Cancer* **49**, 5–10.
Woodruff, M. F. A., Ansell, J. D., Hodson, B. A., and Potts, R. B. (1986a). Oligoclonal tumours. *Int. J. Cancer* **38**, 747–51.
Woodruff, M. F. A., Hodson, B. A., and Deane, D. L. (1986b). The effect of passage *in vitro* and *in vivo* on the properties of murine fibrosarcomas. *Br. J. Cancer* **54**, 621–9.
Wright, S. (1982). The shifting balance theory and macroevolution. *Annual Rev. Genet.* **16**, 1–19.
Wyke, J. A. and Weiss, R. A. (1984). The contribution of tumour viruses to human and experimental oncology. *Cancer Surveys* **3**, 1–24.
Yamasaki, H. (1986). Cell–cell interaction and carcinogenesis. *Toxicol. Pathol.* **14**, 363–9.
Yamasaki, H. and Katoh, F. (1988). Further evidence for the involvement of gap-junctional intercellular communication in induction and maintenance of transformed foci. *Cancer Res.* **48**, 3490–5.
Yamasaki, H., Hollstein, M., Mesnil, M., Martel, N., and Aguelon, A. M. (1987). Selective lack of intercellular communication between transformed and non-transformed cells as a common property for chemical and oncogene transformation of BALB/c 3T3 cells. *Cancer Res.* **47**, 5658–64.
Yoshida, M., Seiki, M., Yamaguchi, K., and Takatsuki, K. (1984). Monoclonal integration of human T-cell leukemia provirus in all primary tumors of adult T-cell leukemia

suggests causative role of human T-cell leukemia virus in the disease. *Proc. Natl. Acad. Sci. USA* **81**, 2534–7.

Yron, I., Wood, T. A., Spiess, P. J., and Rosenberg, S. A. (1980). *In vitro* growth of murine T cells. V. The isolation and growth of lymphoid cells infiltrating syngeneic solid tumors. *J. Immunol.* **125**, 238–45.

Zimmermann, A. and Keller, H. U. (1987). Locomotion of tumor cells as an element of invasion and metastasis. *Biomedicine and Pharmacotherapy* **41**, 337–44.

Index

acoustic neuroma 22
adenopolyposis (polyposis) of colon 1, 22, 34, 40, 74
adjuvant therapy *see* chemotherapy
adoptive immunization 100-3
alkylating agents 80, 84
aminoglutethamide 88, 89
aminomethyltrimethyl psoralen 95
anaplastic cells 9
angioneogenesis factor 3, 104
aniridia 21
anti-angioneogenesis factor 104
antibodies
  anti-ideotype 96
  antitumour 96
  monoclonal 6, 44, 85, 86, 96
  natural 50
  production 42
  see immunochemotherapy; immunotherapy
antigens, tumour (TAA, TATA) 2, 6, 29, 32-3, 35, 94
antimetabolites 80, 82, 84, 105
anti-oncogene *see* tumour suppressor gene
asocial cells 24
athymic mice 51, 59, 63
autochthonous tumours, definition 4, 8
autocrine stimulation 14
autonomy, concept of 48
autoradiography 25
azacytidine 94
azathioprine 67

B-cell (B-lymphocyte) 17, 42
  growth factor (BCGF) 17 *see also* interleukins
  lymphoma *see* lymphoma
BCG 100
basic research 107
benign tumours 1, 2, 54
biological response modifiers 100
biotherapy 93
bladder, carcinoma 22, 53, 63, 73
bleomycin 92

blood transfusion 37, 73
bone marrow
  stimulation 103
  transplantation 81, 84-6, 99
breast cancer 1, 22, 65, 75-8, 82, 87, 88
  mortality 76-8
  screening 75
Burkitt's lymphoma (BL) 17, 34, 48-9, 80

cancer
  curability 71
  defence (homeostatic) mechanisms 1, 24, 25, 48
  definition 2
  early diagnosis 75
  genetic basis 19
  treatment *see* treatment
  see also carcinoma; neoplasm; tumour
carcinogen(s)
  chemical 24, 26, 32, 33
  definition 10
carcinogenesis 2, 9-27, 28, 37-40, 44, 52
carcinoma 2, 18-19, 22, 25, 32, 34, 40, 50, 51, 53, 62, 63, 65, 69, 72, 75-8, 82, 87, 88, 103
  basal cell 34, 55, 68
  definition 2
  in situ 75
  murine 34, 50, 51, 62, 67
  squamous cell 25
  see also tumour; *and particular sites*
cell(s)
  cycle 3-4, 6, 33, 47, 87, 105
  dormant *see* dormancy
  dynamic heterogeneity 41ff., 86, 91, 107
  hybridization (fusion) 20, 22, 43
  hypoxic 92
  interaction 43, 49ff.
  lines 6, 12, 15, 16, 17, 26, 34, 63
  lineages 5, 6, 9, 47
  migratory activity 54
  number in body 24
  sorting 5, 6

cell(s)—(contd.)
  stem 4, 24–5, 46–7, 99
  types of in tumours 1ff., 44ff.
  see also B-cell(s); T-cell
cervix, carcinoma 32, 34, 40, 72, 75
chemotherapy 27, 79ff., 92
  adjuvant 80, 82–4, 87, 88, 105–6
  complications 80–1, 87–8
  pre-operative (neo-adjuvant) 84, 105
  regional perfusion 81–2
chimeras 27, 29, 33, 34
chorioncarcinoma 69, 80, 99
chromosome(s)
  aberrations 16
  breakage 45, 59
  deletions 17, 20, 21, 22, 31, 42
  double minute 46
  non-dysjunction 45
  Philadelphia 5, 13, 17, 31
  selective loss 4, 43
  t(X:11) 23
  transfer 20, 22, 25
  translocation 13, 17, 18, 31, 42, 45
  see also karyotypic abnormalities;
    polyploidy; X-chromosome
clinical therapeutic trials see treatment
clonality of tumours 28ff., 39
clone
  definition 28
  number of in tumours 25–6
  properties 50–2
  spatial distribution in tumours 36–7
cloning
  of genes 21
  of tumour cells 7, 49ff.
colony stimulating factor 103
colorectal carcinoma 18–19, 22, 34, 40, 53,
    58, 82, 83, 99, 102, 103
contact inhibition 8, 54
corticosteroids 104 see also prednisone and
    prednisolone
Corynebacterium parvum 100
cyclophosphamide 51, 102
cyclosporin (A) 59
cytocidal and cytotoxic drugs 80
cytosine arabinoside 27

De Gaulle, Charles 71
defence mechanisms 24, 25
development, normal 42–4, 46
dexamethasone 26
dicoumarin 92
differentiation induction 20, 25–6, 27, 93, 103
distamycin 96
DNA
  binding agents 33, 95, 96, 105

  cloned 44
  methylation 43, 94
  probes 21
  rearrangement 45
  repair 24, 45, 86, 87, 92, 93, 105
  replication (synthesis) 24, 42, 89, 105
  single-stranded 95
  staining for 5, 6
  viruses 32
  see also recombinant DNA technology
doomed cells 4
dormancy 64–9
dose-response curves 84
doxorubicin 103
drug–antibody conjugates
  see immunochemotherapy
drug resistance 86–7, 92, 94
Dryden, John 2
Dukes' grading of colorectal cancer 83
dynamic heterogeneity of tumour cells 7, 41ff.,
    55, 84, 86, 91, 107

echinomycin 96
ecology of tumour cell populations 41
embryonic induction 43
emerogene see tumour suppressor gene
endocrinological procedures 88–91
environmental factors in carcinogenesis 23–5
enzyme–antibody conjugates
  see immunochemotherapy
epicortisol 104
epigenetic mechanisms 42, 46
epidermal growth factor 14, 103–4
Epstein–Barr virus 11, 17
  see also Burkitt's lymphoma
esterase D 21
Ewing's sarcoma 79
exfoliative cytology 75
extracellular matrix 2, 43–4, 46, 54, 55

fast neutron therapy 73
fibrosarcoma (rodent) 32, 35, 36, 37, 49, 50,
    51, 52, 61, 68
first order kinetics 82
flow cytometry 47

G_0 cells see cell, cycle
ganglioneuroma 1, 27, 53
gap junctions 4, 17, 20, 26, 43, 46
Garrick, David v
gene(s)
  activation 46
  amplification 13–14, 45, 46, 48, 86–7
  expression, regulation 43, 93, 94, 95

housekeeping 42
modulator 11, 23
multiplication 45, 46
regulatory 19
transforming 19
tumour suppressor *see* tumour suppressor gene
see also Ig genes; oncogenes
glioma 79
glucose-6-phosphate dehydrogenase 29, 30, 34
goitre 1
Gompertz function 69
graft versus host disease (reaction) 85
growth factors 12, 14, 17, 25-6, 38, 46, 60, 93, 94, 103
see also named growth factors
growth factor receptors 12, 14, 62, 94

Halsted's operation 76
HeLa cells 23
hepatic artery ligation and perfusion 82
histocompatibility
   HLA system 6, 85
   minor antigens 85
Hodgkin's lymphoma 79, 80, 87
Hoechst 33342 dye 5
hormone dependence 48
host reaction to tumours 7, 8, 25, 47-9, 68
   see also defence mechanisms
hyperbaric oxygen 73
hyperplasia 1

ideosomatic predators 24
Ig genes 29, 31, 35
immortalized cells 8, 12, 17, 24
immunochemotherapy 96-9
immunoglobulins 29, 35, 98
immunotherapy 99ff.
insertional mutagenesis 41, 42
insulin 26
intercalating agents 95
interferon 100
interleukins 100-3

Kaposi's sarcoma 79
karyotypic (chromosomal) abnormalities 5
karyotypic (chromosomal) markers 29, 31
kidney
   carcinoma 66-8, 101, 102
   transplants 68

lactogenic hormone 43-4
larynx, carinoma 53, 72

laser beam therapy 72, 73
leucophoresis 102
leukaemia
   acute lymphoblastic 80, 88
   myeloid 17, 25, 26, 27, 31, 34, 85, 103
   other 45, 87, 100
Lewis lung carcinoma 58
lexitropsins 96
life tables 77
lip, mouth, and tongue, carcinomas 34, 72
liver (hepatic) carcinoma 32, 40, 53
long terminal repeat 11
lung, carcinoma 17-18, 21, 45, 53, 80, 103
luteinizing hormone release hormone 89
lymphocytes, tumour infiltrating 100-3
   see also B-cell(s); T-cell
lymphokine activated killer cells 100-3
lymph node(s)
   enlarged, significance 70
   metastases 56, 70
lymphoma
   human 34, 35, 37
   murine 35, 39, 50, 94
   see also Burkitt's lymphoma, Hodgkin's lymphoma

macrophage 3, 8, 101
MHC molecules 61, 62, 94
malignant neoplasm *see* neoplasm; tumour
mammary carcinoma
   human *see* breast cancer
   murine 50, 51, 62, 92-3
mammography 76
mastectomy 65, 70, 76-8
matagens 95
melanoma
   human 53, 65-7, 69, 75, 78, 82, 101, 102
   murine 49, 50, 58, 61
melphalan 103
meningioma 1
metaplasia 9
metastasis(es) 18, 54-70, 84, 103, 104
   analysis, in animals 58-60
   artificial 49, 58-60
   occult residual 104-5
   regression 69, 73
   treatment 101, 102
   see also dormancy
methotrexate 87
methylcholanthrene 32, 39
misonidazole 73, 92
mitomycin C 67, 92
monoclonal antibodies *see* antibodies
monoclonality of tumours, evidence 29ff., 40
multiple endocrine neoplasia 22
mutagens, environmental 45

mutation 19, 24, 41-2, 44-6
myc oncogenes 12, 14, 15, 17, 18, 19, 35, 39, 45, 48, 63

nasopharynx, carcinoma 32
neoplasm
  benign 2
  definition 1, 2
  malignant 2, 47, 48
  see also carcinoma; sarcoma; tumour
neuroblastoma 27, 45, 48, 53, 69, 80
neurofibroma 1, 22, 33
NIH 3T3 cell line 12, 15, 16, 17, 26, 63
NK cells 8, 48, 100
nucleases 95
nude mice *see* athymic mice

occult residual cancer 83-104
oestrogen receptors 89-91
oestradiol 89
oligonucleosides 94-5
oligonucleotides 94-5
oligopeptides 95-6
oncogene(s) 10, 11-19, 45, 48, 63, 94
  activation 48
  amplification 63
  expression 63
  influence on metastasis 63
  see also myc oncogenes; *ras* oncogenes
oncogenic package 10, 37
orchidectomy 89
ornithine carbamoyl transferase 37
osteo(genic) sarcoma 21, 22, 31, 69, 78, 83
ovary
  carcinoma 69, 87, 102
  irradiation 88
  removal (oophorectomy) 88, 89

P-glycoprotein 86
pancreas, carcinoma 53
para-immunological mechanisms 8, 25
phenotypic diversity, genetic basis 41-7
Philadelphia chromosome 5, 13, 17, 31
phosphoglycerate kinase (PGK-1) 29, 30, 32, 35, 36, 50, 51
plasma cell(s) 42
  leukaemia 45
plasmacytoma 45
platelet derived growth factor 14
pleiotropic drug resistance 86, 92
pleoclonal (polyclonal) tumours 32, 35-7
polynucleotides, synthetic 100
polyoma antigen 15
  virus 11, 26

polyploidy 45
polyposis *see* adenopolyposis of colon
prednisone and prednisolone 67, 88
preneoplastic (part way) cells 4, 16
procarcinogens 33
progression, tumour 7, 10, 14, 45, 46, 52-3, 69
promoters 9
prostate, carcinoma 73, 88, 89
protein kinases 12
proto-oncogenes 11, 12ff.
  normal functions 12, 13, 94
  role in carcinogenesis 13, 14ff.

radiosensitivity of tumours 72
radiosensitizing agents 73
radiotherapy 72, 73, 76-80, 85, 87-8
radon 73
ras oncogenes 12, 14, 15, 17, 18, 19, 20, 22, 26, 45, 63, 64
razoxane 83
recombinant DNA technology 93, 102, 103
regression, spontaneous
  of metastases 69, 73
  of primary tumours 52-3
restriction fragment length polymorphism 29, 30
retinoblastoma 20-1, 31, 39, 45, 87
  gene 21, 22
retinoic acid 26, 45, 48
retrovirus *see* virus
revertant cells 3, 4, 20
RNA 18, 33, 95 *see also* virus
RNase 18
Rous sarcoma virus 9, 11

SV40 virus 11
sarcoma 33, 53
  see also fibrosarcoma; osteo(genic) sarcoma
screening for cancer 75-6
second mesages 14, 15, 16
sister chromatid exchange 45
skin
  carcinomas 9, 24, 25, 55, 68, 72
  grafts 68
small bowel, carcinoma 24
small cell carcinoma of lung 17-18, 45, 103
Southern blotting 32
specific growth rate 69
stathmokinetic agents 47
stilboestrol 88, 89
stomach, carcinoma 53
surgery, role in cancer 72, 73, 74
synkavit 73

T-cell (T lymphocyte) 42, 85, 86, 100
  deficient mice 51, 59, 63
  growth factor *see* interleukins
  neoplasms 32
  receptor 29, 31, 32
tamoxifen 88, 89, 90
teratocarcinoma 26, 27
testis
  intra-abdominal 74
  tumours 69
Tetrahydro S 104
thoracic duct drainage 68
thymidine, tritiated 25, 47
thyroid tumours 1, 53, 88, 89
thyroidectomy 89
thyrotrophic hormone 89
thyroxin 89
totipotent cells 42
transcription 14
transduction 14
transfection 15, 63, 64
transferrin 44
transformation, neoplastic 3, 4, 8, 9, 10, 11ff., 38
  environmental factors affecting 23-5
  reversibility 25-7
transformed cells 3, 4, 8
transgenic mice 19, 35
translation, blocking of 95
transplantation
  bone marrow *see* bone marrow
  cell 42
  nuclei 42
  organ 68, 73
treatment
  clinical trials 71, 76-8, 89-90, 101, 102, 106
  current methods and their limitations 71-91
  results 71-2
  ways forward 92-106
trichoepithelioma 33
tumorigenicity 8, 25
tumour(s) *see also* neoplasm
  angioneogenesis factor 3, 104
  antigens *see* antigens
  autochthonous 4, 8
  benign 1-2, 54
  biclonal 38-9, 52
  conglomerate 28

definition 1, 2
experimental 5
extracellular matrix 2, 43
hereditary 19
malignant 1, 54
monoclonal 34
murine 32, 34-7, 49-52, 62, 68, 69
necrosis factor 101
pleoclonal (polyclonal) 33, 35-7
progression *see* progression
spontaneous 5
structure 2ff.
see also sites and names of tumours
tumour cell(s) *see also* cell(s)
  cloning 7
  ecology 7, 41
  generation of diversity 44ff.
  markers 6-7
tumour suppressor gene 4, 11, 17, 19-23, 25, 26, 93
twin donors 85

ulcers, carcinomatous 25
ultraviolet radiation 24, 95

virus (viral)
  DNA 32
  integration sites 29, 32
  mouse mammary tumour 51
  oncogenes 11
  retrovirus 10, 11, 45
  Rous sarcoma 9, 11
  T-cell lymphoma 32

Walker rat carcinoma 58, 67
Werner's syndrome 40
Wilms' tumour 20, 21, 23, 31, 34, 39, 80, 87

X-chromosome 23, 29-30, 31, 36
  inactivation 29-30, 36
X-irradiation 26
X-linked markers 28, 29, 31
X-ray therapy 72, 73
xeroderma pigmentosum 24

RC
268.5
.W62
1990

32.50